IMAGES
of America

FORT BLISS

ON THE COVER: Trick riders demonstrate their abilities on Smith Bliss Field in 1922. These were likely 5th Cavalry soldiers showing off their riding skills. The 1893 infantry barracks and the 1915 cavalry barracks are in the background. The picture was developed by the Hill Photo Shop, which existed in the small community of Lynchville, northwest of the post. The photograph is now a part of the Fort Bliss Museum collection. (1st Armored Division and Fort Bliss Museum.)

IMAGES
of America

FORT BLISS

John A. Hamilton

ARCADIA
PUBLISHING

ISBN 978-1-4671-2915-2

Published by Arcadia Publishing
Charleston, South Carolina

Printed in the United States of America

Library of Congress Control Number: 2017961198

For all general information, please contact Arcadia Publishing:
Telephone 843-853-2070
Fax 843-853-0044
E-mail sales@arcadiapublishing.com
For customer service and orders:
Toll-Free 1-888-313-2665

Visit us on the Internet at www.arcadiapublishing.com

This book is dedicated to my dear wife, Vicki, and to the soldiers and civilians whose hard work makes the post what it is today.

Contents

ACKNOWLEDGMENTS

The author gratefully acknowledges the 1st Armored Division and Fort Bliss Museum for access to the extensive photograph collection. This collection was part of the Otis Aultman collection, much of which was in the hands of Comdr. Millard McKinney (Ret.) until his death. Some of this went to the Border Heritage Collection in the El Paso Public Library, some to the library of the University of Texas El Paso, and some to the museum on Fort Bliss. There is much to be learned in these sources. The museum on Fort Bliss also provided access to many official US Army photographs.

Unless otherwise indicated, all photographs are from the 1st Armored Division and Fort Bliss Museum.

INTRODUCTION

El Paso, Texas, is a thriving and diverse community of around 800,000 at this writing. It is one of the largest cities in Texas, and it is also one of the farthest removed from other major Texas cities. Built around the nexus of major highways and railroads, it has been and still is a major center of commerce as well as a major gateway to and from Mexico to the south. Cíudad Juárez, across the Rio Grande, is another flourishing community and a partner city to El Paso. The dividing line is the once-mighty Rio Grande, now controlled by a series of dams and watercourses in most seasons. But in an earlier time, the river's Great Pass of the North was important, and the Rio Grande was more of a barrier as well as a source of water for residents and travelers.

The El Paso area has a rich and varied history, and a large part of that history is the enduring presence of military forces. Soldiers at the pass of the Rio Grande were there from the earliest days to provide protection to the civilian community, to secure the southern border, and to train for war. The US Army forces in and around El Paso have contributed to the very rich history of the borderlands over the 168 years that they have been here.

In writing a military history of El Paso, one cannot forget the presence of Spanish and Mexican forces that were in the area for many years. After the arrival of the Spanish conquistador Don Juan de Oñate in 1598, Spanish soldiers, then Mexican soldiers, occupied the Great Pass of the North. During that conflict, Missouri volunteers, under the command of Col. Alexander Doniphan, fought an engagement with Mexican troops on Christmas Day in 1846 at a place called Brazito, north of the pass of the Rio Grande. Doniphan's forces were victorious, moving into the old village of El Paso del Norte on the Mexican side of the river. There they rested and reprovisioned before moving on into Mexico to join Gen. William Worth's American forces at Saltillo.

Before the town of El Paso existed, and before the Mexican village of El Paso del Norte became Cíudad Juárez, there was a US Army post here. In 1848, there were a few Americans living on the north side of the Rio Grande. The Treaty of Guadalupe Hidalgo ended the Mexican-American War and led to the Army's establishment of a post opposite the Mexican town, across the river. The river became the new international boundary. The Army established the Post Opposite El Paso in 1849, and this would eventually become Fort Bliss. The post closed and reopened twice, expanded and contracted many times, and shifted locations six times in its long history.

The history of Fort Bliss can be divided into several distinct periods. First, the establishment of the post and its existence prior to the American Civil War was primarily a period of exploration and conflict with the Indians. The initial Post Opposite El Paso was established in 1849, but it was closed in 1851, with the troops moving to a post called Fort Fillmore near Old Mesilla, New Mexico. The El Paso post was reestablished in 1853, on land leased from a local trader and rancher, James Wiley Magoffin. It was this post that was named for Lt. Col. William Wallace Smith Bliss, who had served as Gen. Zachary Taylor's adjutant general in the Mexican-American War. It was established in the community known as Magoffinsville. This period lasted until 1868 and saw continued conflict with the local Native American tribes. This post also saw the advance and withdrawal of the Confederacy, whose volunteers hoisted the rebel flag and held the post until the defeat and retirement of the Sibley Brigade from New Mexico back to San Antonio. The Union flag was restored in 1862, and after that, regular US Army troops garrisoned the post once again. Aside from the Civil War period, most of the troops stationed at Fort Bliss were infantry augmented with small numbers of horse cavalry and some field guns.

When the Rio Grande washed the old post at Magoffinsville away in 1868, the Army found a new home for Fort Bliss at Camp Concordia, almost exactly where the junction of Highway 54 and Interstate Highway 10 is today. This post was closed in 1877, then reopened in what is now downtown El Paso, around San Jacinto Plaza. In 1879, the post was shifted to land purchased next to entrepreneur Simeon Hart's gristmill on what is now Paisano Drive, west of downtown El Paso. When the railroad finally came to El Paso in 1881, the old post at Hart's Mill was crisscrossed with railroad tracks. This forced the post to move once again.

The next major historical period for the post was from 1893 until 1940. After some negotiations between the El Paso city fathers and the War Department, the Army moved the post from Hart's Mill to a broad plain about five miles northeast of the town. During this period, the Army experienced the Spanish-American War, the Mexican Revolution, and World War I. This was a period of dramatic change, with the post's garrison shifting from infantry to horse cavalry. The cavalry was necessary to patrol the border, especially during the Mexican Revolution.

Fort Bliss's next historical period runs from 1940 until 2009, with the departure of the 1st Cavalry Division to fight as infantry in the Southwest Pacific and the arrival of a new branch of the Army, the antiaircraft artillery. This has been the post's longest period of history in one place. With its clear weather and open airspace, Fort Bliss became the Army's center for antiaircraft training throughout World War II.

The end of the war resulted in a mission change for the post on the mesa. Operation Paperclip in 1946 brought the German V2 rocket scientists to Fort Bliss and White Sands Proving Ground. Subsequently, Fort Bliss became the Army's training and development post for sophisticated air defense missile technology. This was the home of the Army's air defense artillery branch until it moved to Fort Sill, Oklahoma, in 2009.

The last period of history is still unfolding. With its 1.12 million acres of training land and unrestricted airspace, Fort Bliss is the Army's premier maneuver training ground. All weapons in the Army's arsenal can be fired here without restrictions. Ground commanders can maneuver over its vast training land, and air forces can fly training missions in support of ground training without interference.

This brief illustrated history will detail American military history at the Great Pass of the North. It may encourage further study about a post that had modest beginnings and grew into a major Army installation over its long history. This post has a long record of participation in our nation's conflicts, and it has kept a close and productive relationship with the citizens of El Paso during its existence. This work is dedicated to the many soldiers who have served the post and community well throughout its history.

One

Infantry and Adobe
The Frontier Posts

American troops came to the Great Pass of the North during the Mexican-American War, when the Missouri Volunteers, under Col. Alexander Doniphan, arrived in 1846. After winning a battle against Mexican troops at Brazito, Doniphan and his troops moved south into the pass of the Rio Grande. These troops found a thriving community across the river from what is now the city of El Paso. But on the American side of the river, there was only a small collection of ranches and farms. Doniphan's troops moved on into Mexico, and except for a detachment of the 1st US Dragoons, no American troops were in the El Paso area until three years later.

The Mexican-American War ended with the signing of the Treaty of Guadalupe Hidalgo in 1848. The United States had added large portions of the southwest that included what is now Arizona, New Mexico, and California. The requirement to survey the new international border demanded the presence of troop garrisons to establish a military presence and to protect border survey teams. In 1848, the War Department directed the establishment of a military post opposite the Mexican town of El Paso Del Norte, across the Rio Grande from the Great Pass of the North. Accordingly, Bvt. Maj. Jefferson Van Horne, 3rd US Infantry, organized an expedition of six infantry companies to move from San Antonio to the pass of the Rio Grande and establish a military post there. Arriving on September 14, 1849, Van Horne arranged for rented land and buildings and opened the first permanent post on Coon's Ranch.

The post remained there until 1851, when the Army closed it and moved the troops to Fort Fillmore, a new post south of what is now Las Cruces, New Mexico. The post would then move four more times: back to El Paso in 1853 into the community of Magoffinsville, to Camp Concordia in 1868, into rented buildings in downtown El Paso in 1878, and from there into the Hart's Mill location east of downtown in 1879. The post was one of many frontier posts defending against Indian depredations and playing a part in the Civil War in the southwest. In 1854, the Army named the post Fort Bliss after Lt. Col. William Wallace Smith Bliss, West Point math prodigy and principal assistant to Gen. Zachary Taylor, who later became president. The name has endured.

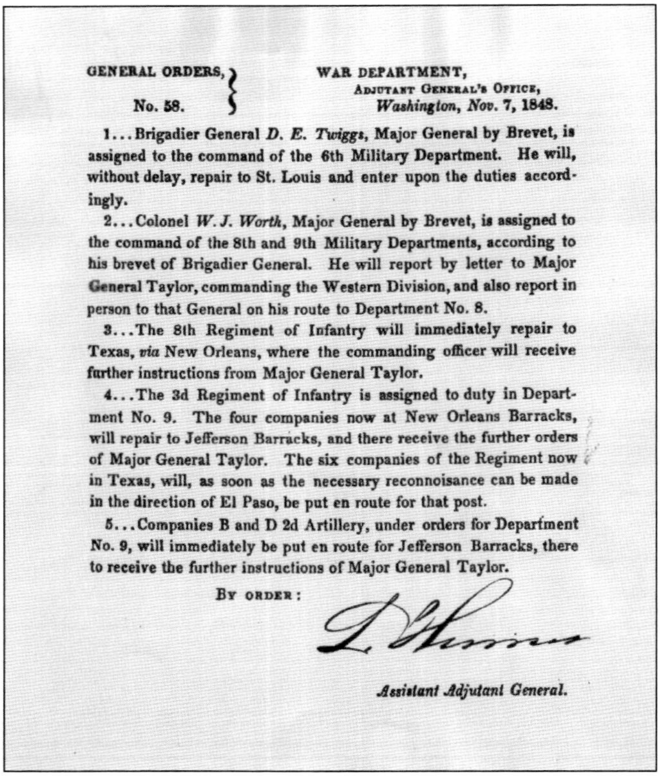

GENERAL ORDERS, No. 58.

WAR DEPARTMENT,
ADJUTANT GENERAL'S OFFICE,
Washington, Nov. 7, 1848.

1...Brigadier General *D. E. Twiggs*, Major General by Brevet, is assigned to the command of the 6th Military Department. He will, without delay, repair to St. Louis and enter upon the duties accordingly.

2...Colonel *W. J. Worth*, Major General by Brevet, is assigned to the command of the 8th and 9th Military Departments, according to his brevet of Brigadier General. He will report by letter to Major General Taylor, commanding the Western Division, and also report in person to that General on his route to Department No. 8.

3...The 8th Regiment of Infantry will immediately repair to Texas, *via* New Orleans, where the commanding officer will receive further instructions from Major General Taylor.

4...The 3d Regiment of Infantry is assigned to duty in Department No. 9. The four companies now at New Orleans Barracks, will repair to Jefferson Barracks, and there receive the further orders of Major General Taylor. The six companies of the Regiment now in Texas, will, as soon as the necessary reconnoisance can be made in the direction of El Paso, be put en route for that post.

5...Companies B and D 2d Artillery, under orders for Department No. 9, will immediately be put en route for Jefferson Barracks, there to receive the further instructions of Major General Taylor.

BY ORDER:

Assistant Adjutant General.

This facsimile of War Department Order No. 58, dated November 7, 1848, directed the 3rd US Infantry Regiment, then at San Antonio, Texas, to send six companies to the pass of the Rio Grande as soon as the necessary reconnaissance of the route was completed. This meant moving on foot about 650 miles through rough terrain where only the barest of roads existed.

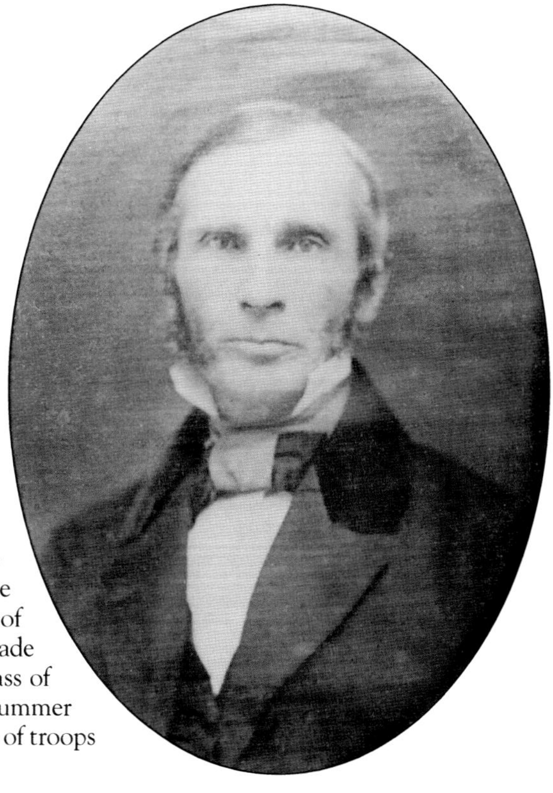

Bvt. Maj. Jefferson Van Horne commanded E Company, 3rd Infantry Regiment in the Mexican-American War. During the war, he was promoted for gallantry at the Battles of Contreras and Churubusco. His infantry made the trek from San Antonio to the Great Pass of the North by walking the 673 miles in the summer of 1849. His was the first permanent garrison of troops in the area. (National Archives.)

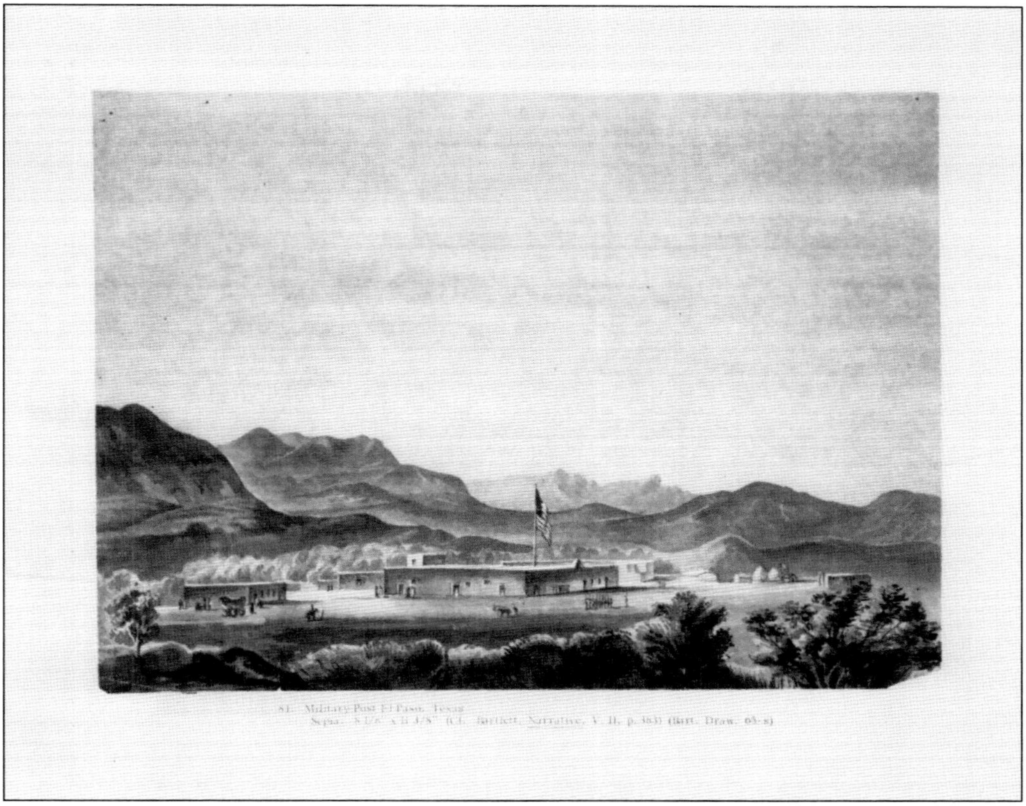

Arriving in the El Paso area on September 8, 1849, the 3rd Infantry Regiment found no town except for a few adobe buildings. Maj. Jefferson Van Horne leased land from local rancher Benjamin Franklin Coons for the post. Four companies of troops moved in, with the remaining two companies and a howitzer battery occupying land at San Elizario. This picture was painted by H.C. Pratt of the first American boundary survey.

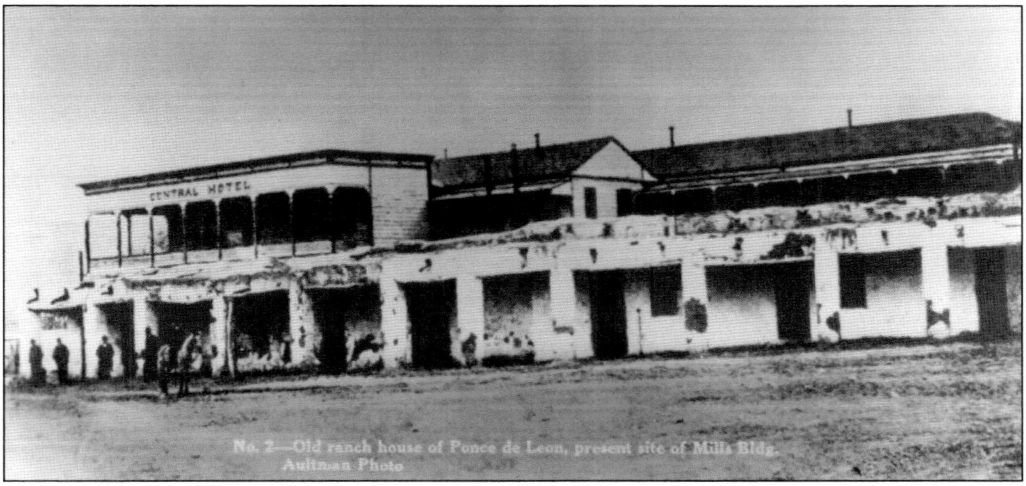

No trace of the original post opposite El Paso exists. This mid-1870s photograph shows a ranch house on the land. This building no longer exists, and downtown El Paso has grown up in its place. El Paso's White House Department Store and the Anson Mills Building would eventually occupy the site.

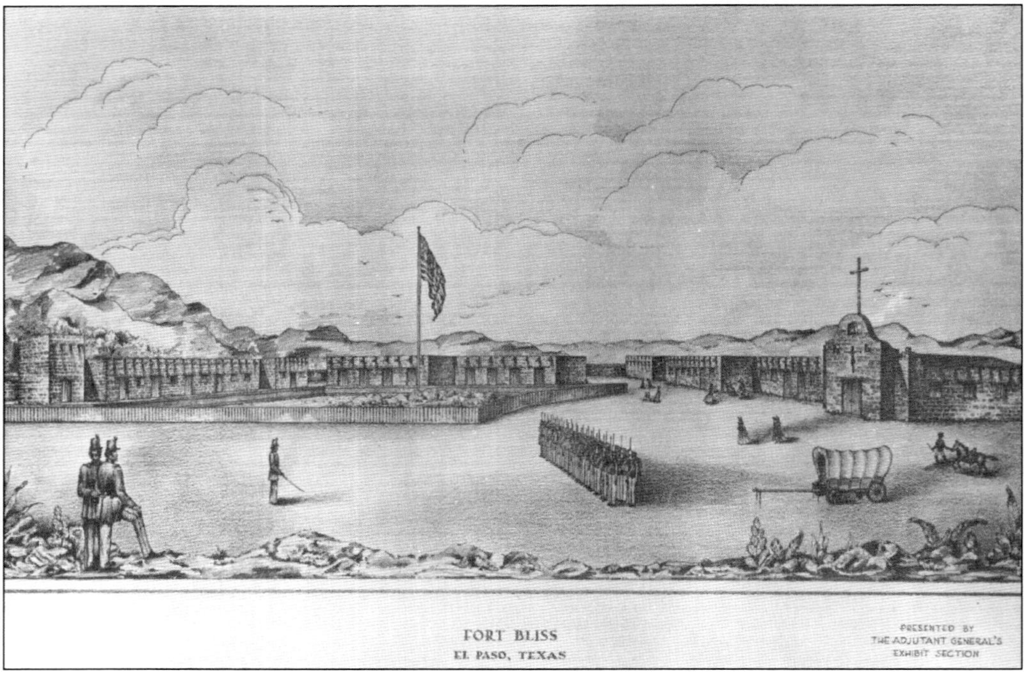

The Army closed the first post in 1851, moving the troops to a post near Mesilla, New Mexico. In January 1854, the Army reestablished the post with troops from the 8th Infantry Regiment. Once again, the Army rented quarters on land owned by James Wiley Magoffin, who became the post trader. This picture depicts the layout of the post, named Fort Bliss for Lt. Col. William Wallace Smith Bliss.

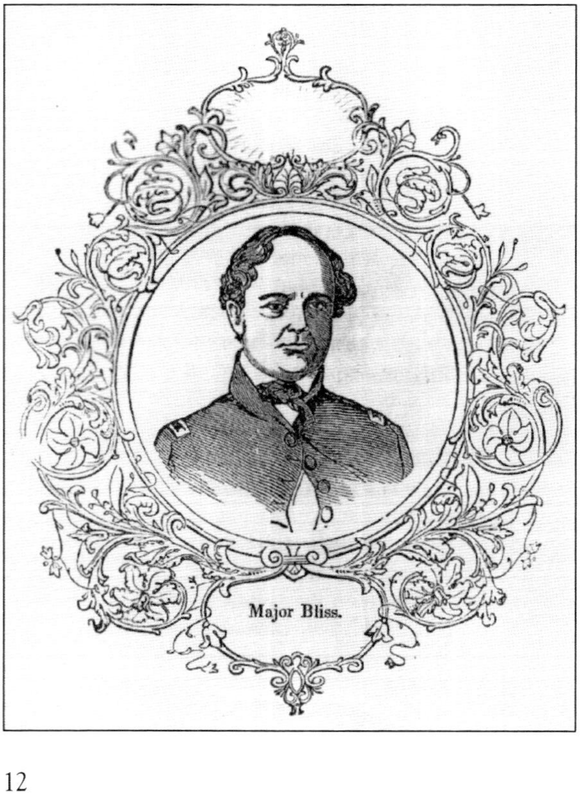

Bvt. Lt. Col. William Wallace Smith Bliss was a mathematics prodigy at West Point. His leadership abilities also made him stand out as a future Army leader. In the Mexican-American War, Gen. Zachary Taylor selected him as his adjutant general and later appointed him his presidential secretary. Bliss perished from yellow fever in Mississippi in 1853, and the Army named the Magoffinsville post for him the following year.

James Wiley Magoffin was a rancher, farmer, trader, and entrepreneur in the El Paso area. He rented quarters to the 8th Infantry troops and served as the post sutler or trader. He supplied beef and other items to Fort Bliss, as the supply line ran 650 miles to San Antonio or 300 miles to Santa Fe. Magoffin would play a part in the post's history until after the Civil War. (Courtesy Magoffin Home, El Paso, Texas.)

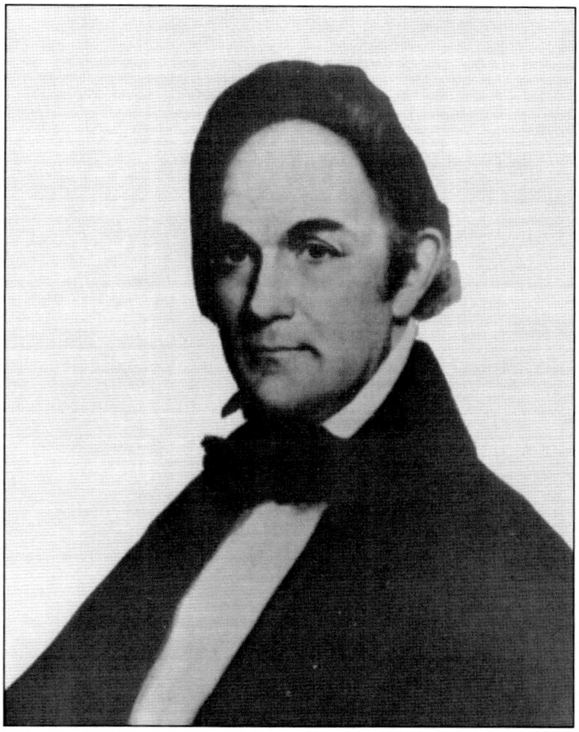

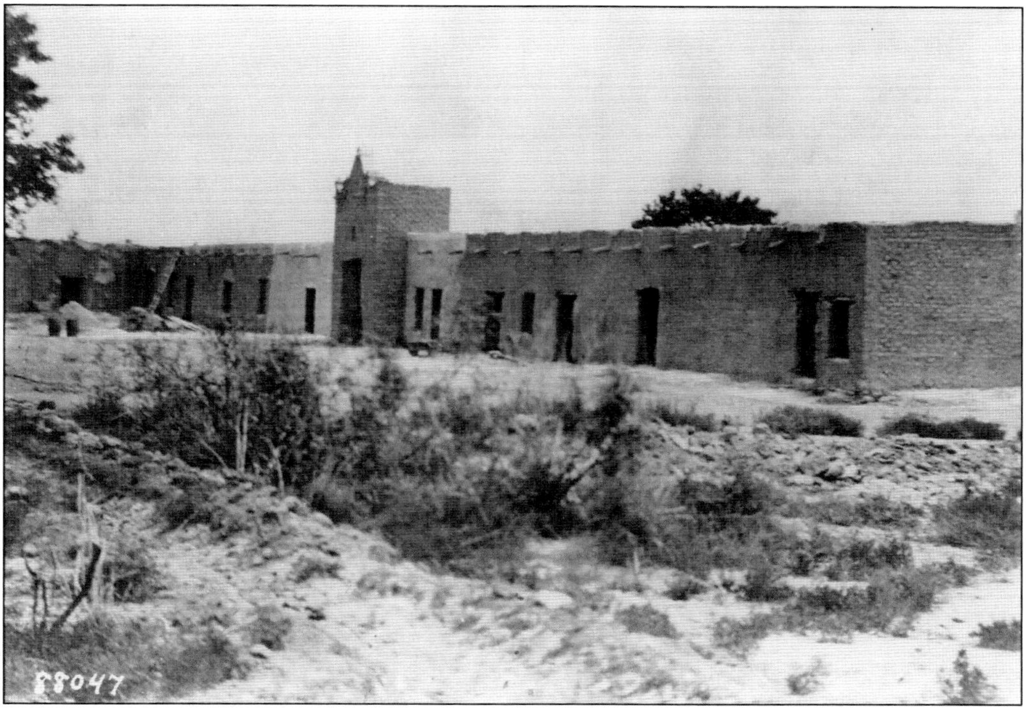

Magoffinsville was built with local materials of adobe and wooden timbers. This is a view of the layout of the storage rooms for provisions, supplies, and equipment. The photograph was taken in 1868, prior to the flood that finally washed the old post away. The rooms were heated by fireplaces in winter and were open in the summertime.

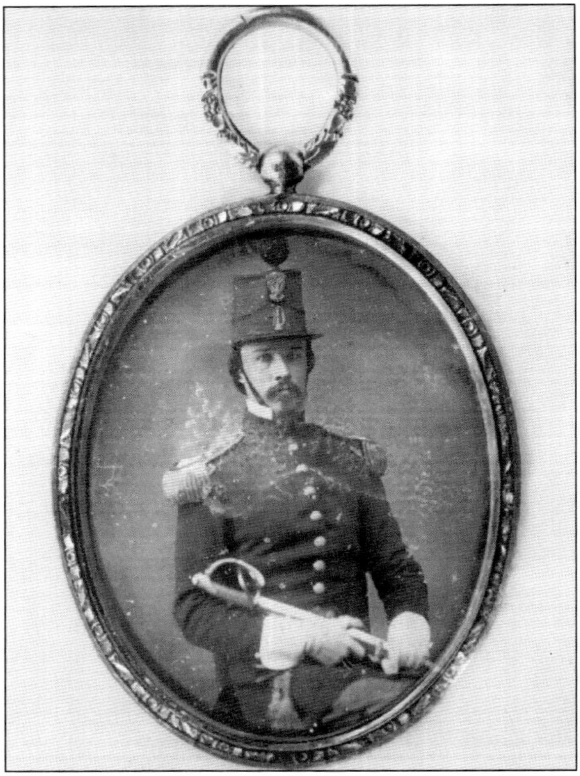

This photograph of Capt. Washington Elliot was taken in 1857 when he was assigned to the Regiment of Mounted Riflemen, the predecessor of the 3rd Cavalry Regiment. Washington commanded Fort Bliss at Magoffinsville between May and October 1859. Command of frontier posts fell to the senior officer on the post. Here, he wears the formal uniform prescribed at the time with the Prince Albert cap.

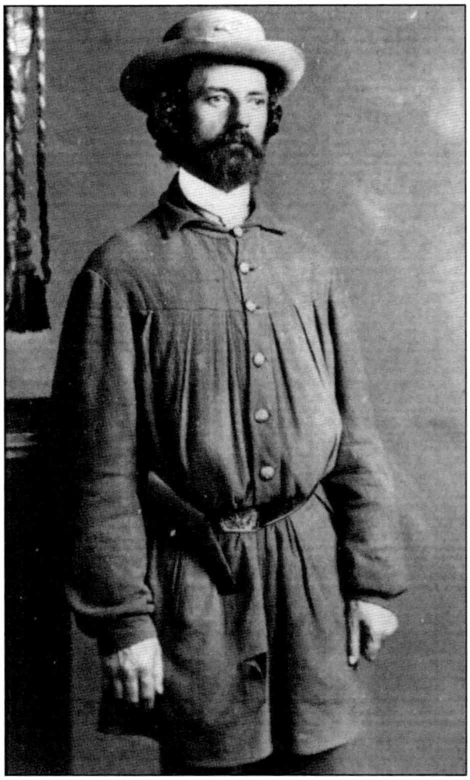

Lt. Henry M. Lazelle was stationed at Fort Bliss with the 8th Infantry Regiment in 1859. He led a detachment of the Regiment of Mounted Riflemen against the Mescalero Apaches that year, fighting an engagement in Dog Canyon, New Mexico, where he was shot through the lungs. He survived to serve in the Civil War and return to command Fort Bliss 34 years later. Here, he is wearing what troops wore on campaign.

When Texas seceded from the Union in 1861, Texas Confederate volunteers moved swiftly to establish claim over former Union army posts. Lt. Col. John Baylor, of the 2nd Texas Mounted Rifles, arrived at Fort Bliss in July 1861. He assumed command of the post and the Territory of Arizona. Baylor commanded forces in the area until he was relieved in 1862 for ordering the extermination of all local Apaches.

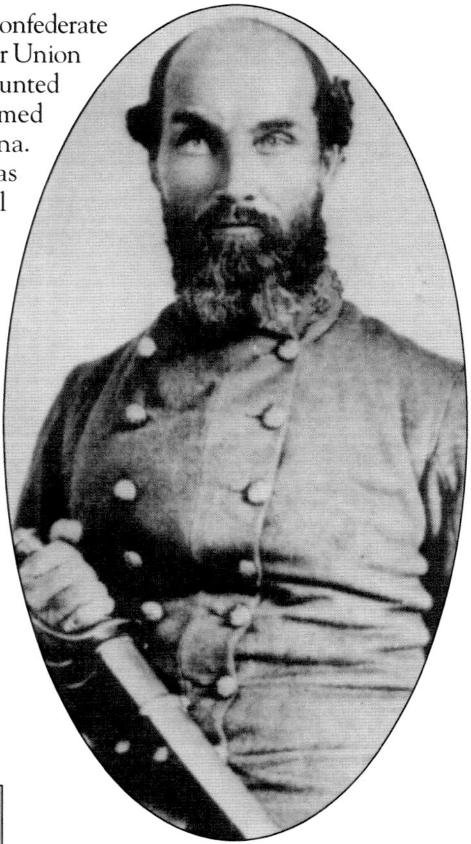

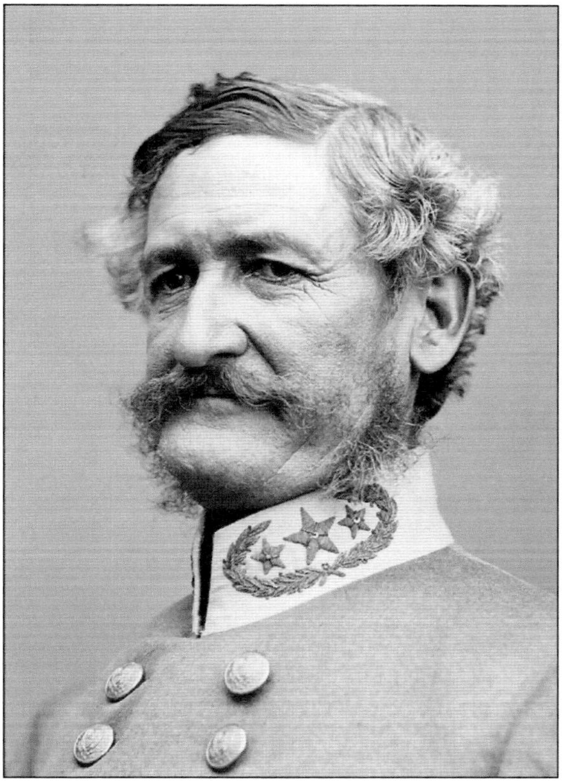

Brig. Gen. Henry Hopkins Sibley arrived at Fort Bliss with his Army of New Mexico in December 1861. He moved on to invade New Mexico in January. Although he won engagements at Valverde and seized Albuquerque and Santa Fe, his forces were defeated at Glorietta Pass, New Mexico, in March 1862. He withdrew through Fort Bliss to San Antonio. Ironically, he was the first general officer to command Fort Bliss. (National Archives.)

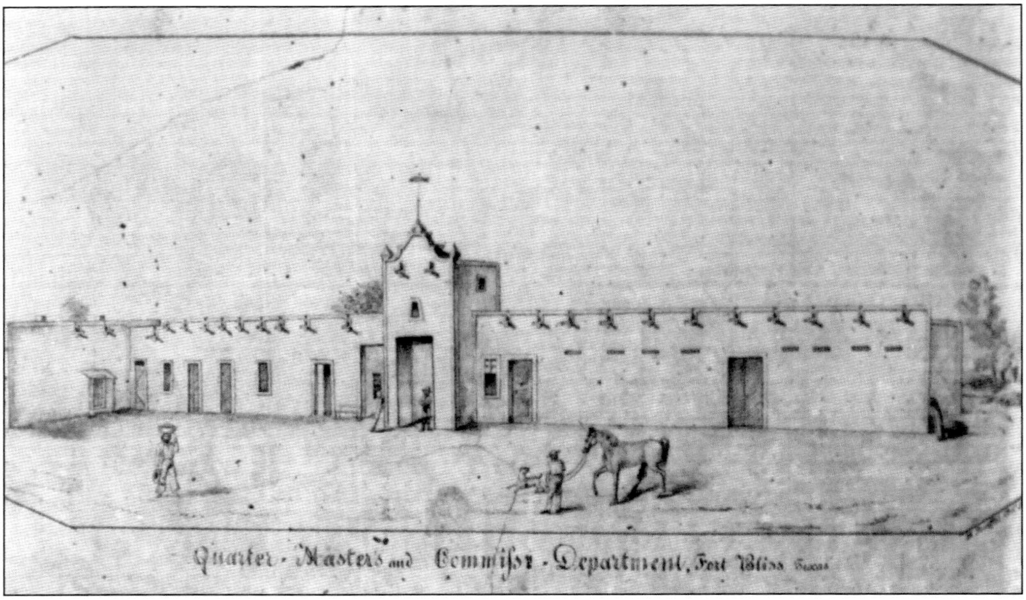

After the Confederates abandoned the Magoffinsville post in 1862, elements of the Union's 1st and 5th California Volunteers occupied the post. They were relieved by regular troops of the 5th Infantry, commanded by Capt. David Brotherton, in 1865. Among Brotherton's troops was Pvt. Herman Stieffel, an amateur artist. He drew this picture of the old commissary and quartermaster buildings, excellent depictions of the actual structures.

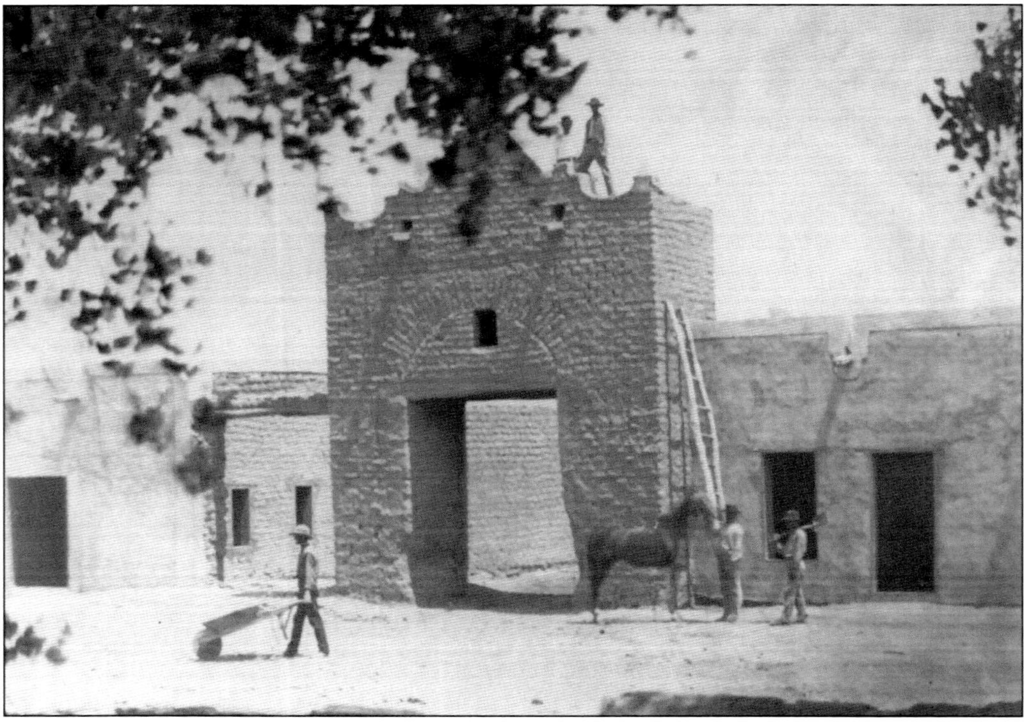

This was the sally port of the Magoffinsville post. It was built of adobe brick and was a remarkable decorative addition to the post. Some have thought it was part of a post chapel, but there is no evidence that it was anything more than a formal entrance.

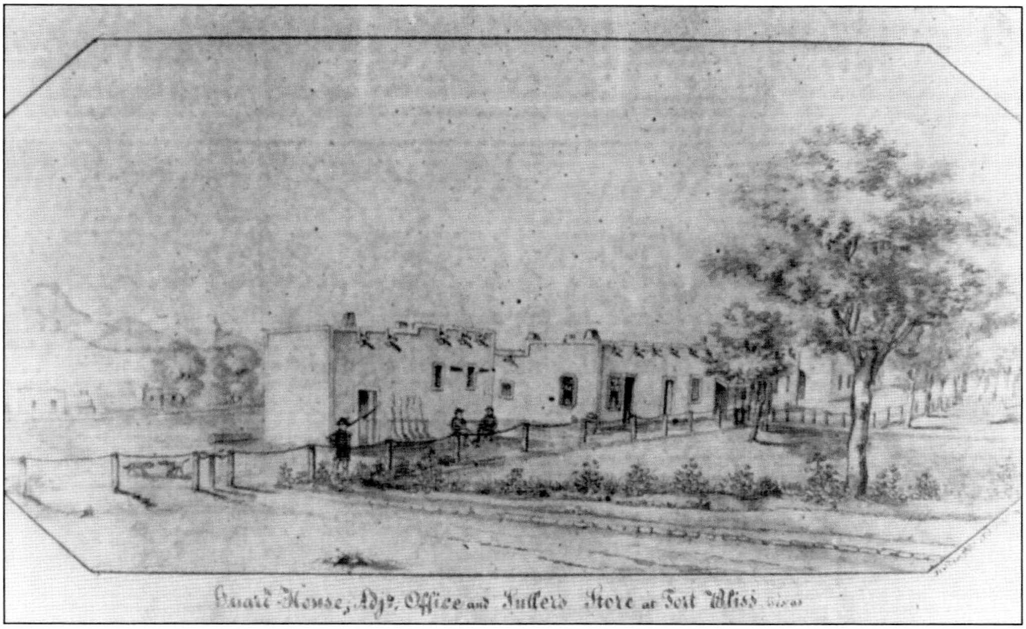

Another of Private Stieffel's drawings shows the Magoffinsville guardhouse, adjutant's office, and sutler's store. James Magoffin was the post sutler, or trader, until the Confederates left in 1862. As a Confederate commissioner, Magoffin supplied the Sibley Brigade and Lt. Col. John Baylor's troops. This cost him his land and buildings after the Civil War. The post adjutant's office served as a headquarters, and the guardhouse held misbehaving soldiers.

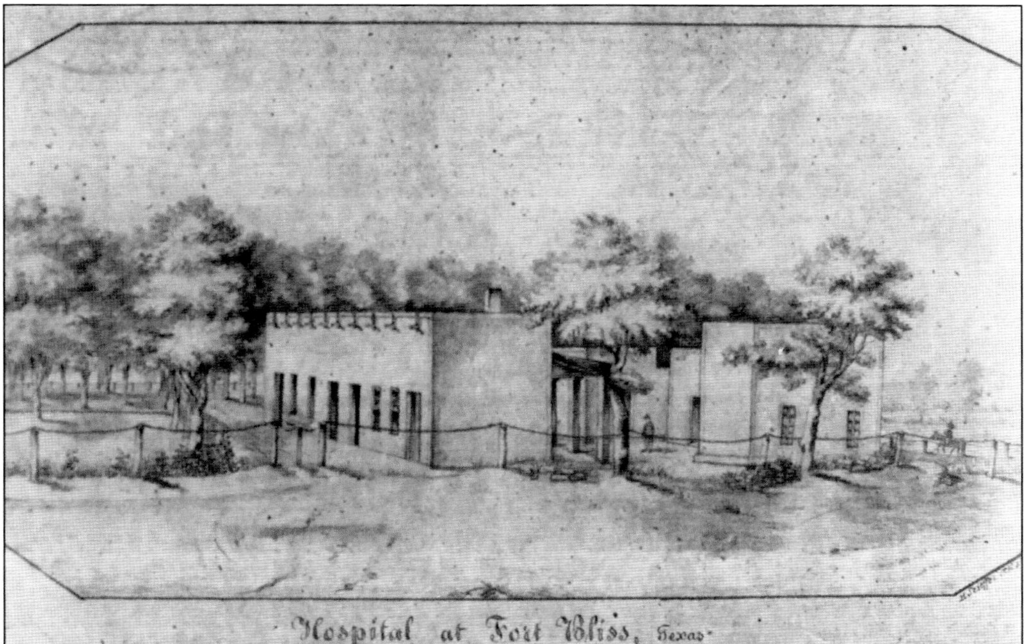

Private Stieffel's drawing of the hospital at Fort Bliss shows another adobe building. Although Stieffel identifies the building as a hospital, there was no specific structure that served as a formal hospital. Likely, this building housed wounded and ill soldiers, starting with the Confederates of the Sibley Brigade that General Sibley left here when the Confederates withdrew.

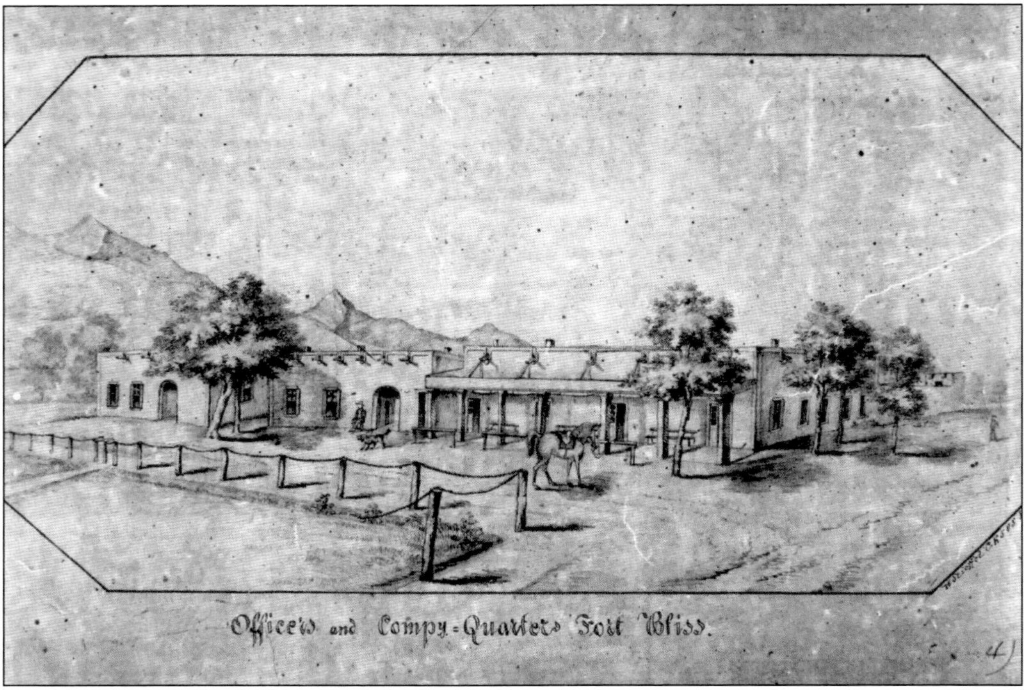

Private Stieffel identified this building as the officers' and company quarters. The troops occupied buildings that had wooden beds and spaces to store personal belongings. Quarters were very crowded and not very comfortable, particularly in winter. There was no mess hall. Troops probably took their meals in the quarters. Water came from the Rio Grande, which meant that troops were exposed to gastrointestinal ailments from organisms in the water.

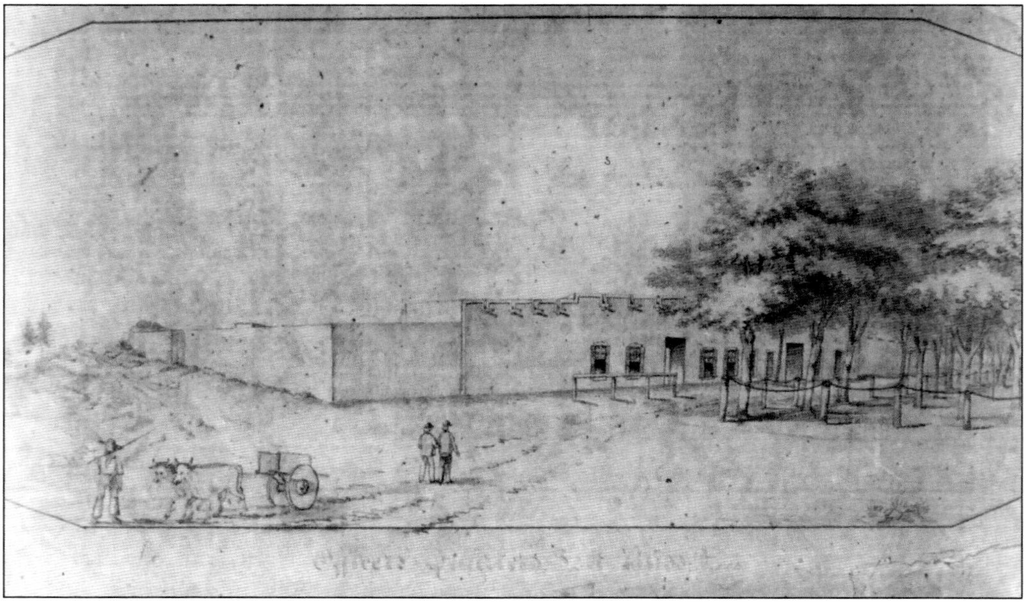

Private Stieffel labeled this building as officers' quarters at Fort Bliss. It probably housed more senior officers. Lydia Lane, married to an officer of the Regiment of Mounted Riflemen, spoke well of their quarters. Their three-room adobe quarters had dirt floors, over which they nailed down wagon canvas and covered that with rugs. With that, she said they were "well fixed."

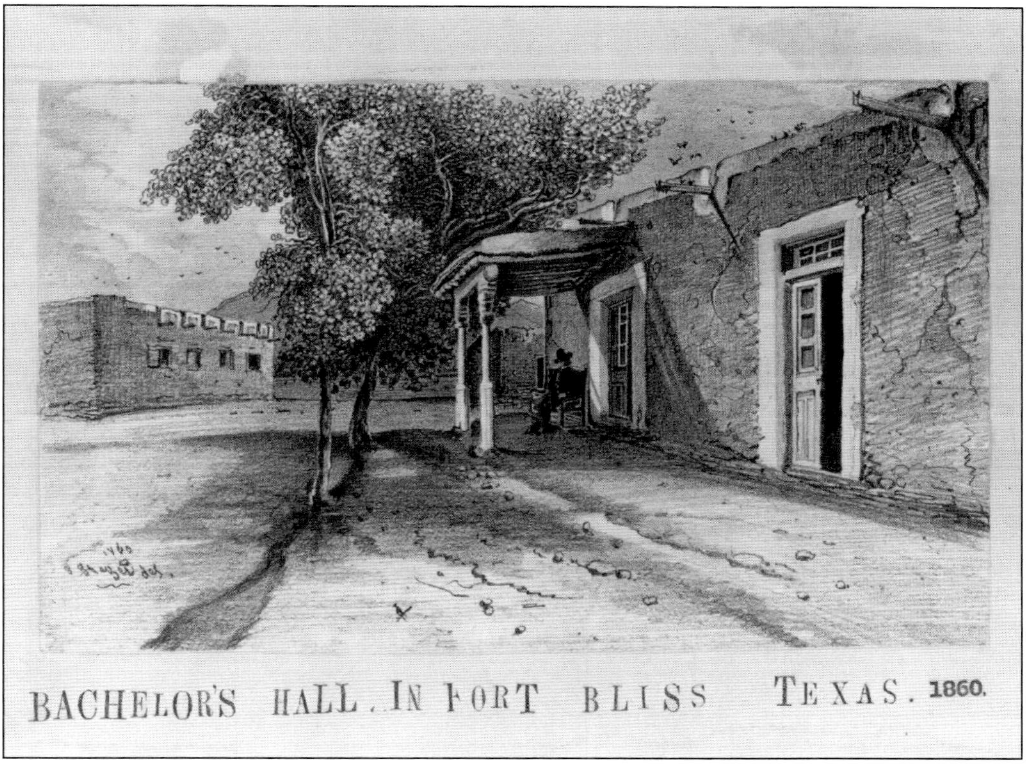

This artist's depiction of the junior officers' quarters at Magoffinsville in 1860 (the year before the secession of Texas from the Union) shows a well-constructed though still quite rustic building. Lydia Lane, spouse of Lieutenant Lane of the Regiment of Mounted Riflemen, spoke highly of their experience at Magoffinsville in 1859. (Original in Magoffin Home, El Paso, Texas.)

This picture was taken around 1868 and may be the same building depicted in the previous drawing. It shows two individuals in civilian clothes lounging outside the building, likely to escape the heat of the interior.

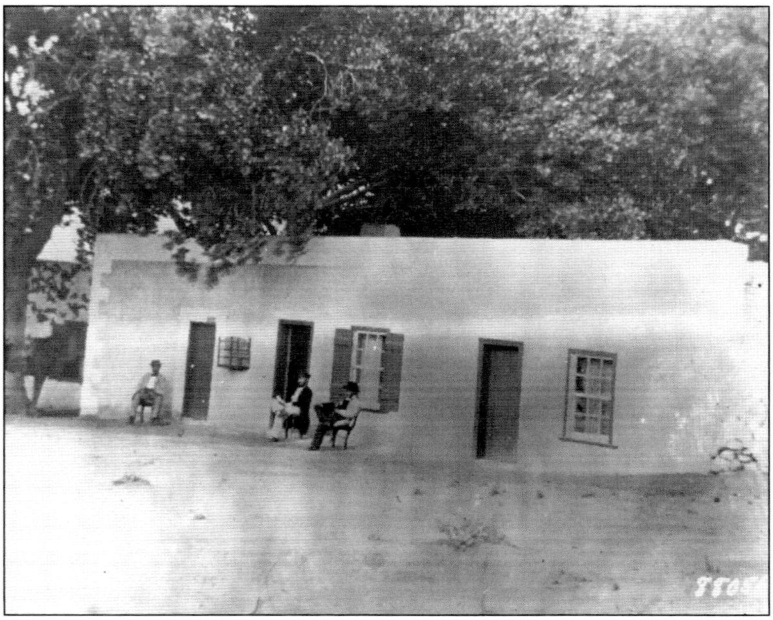

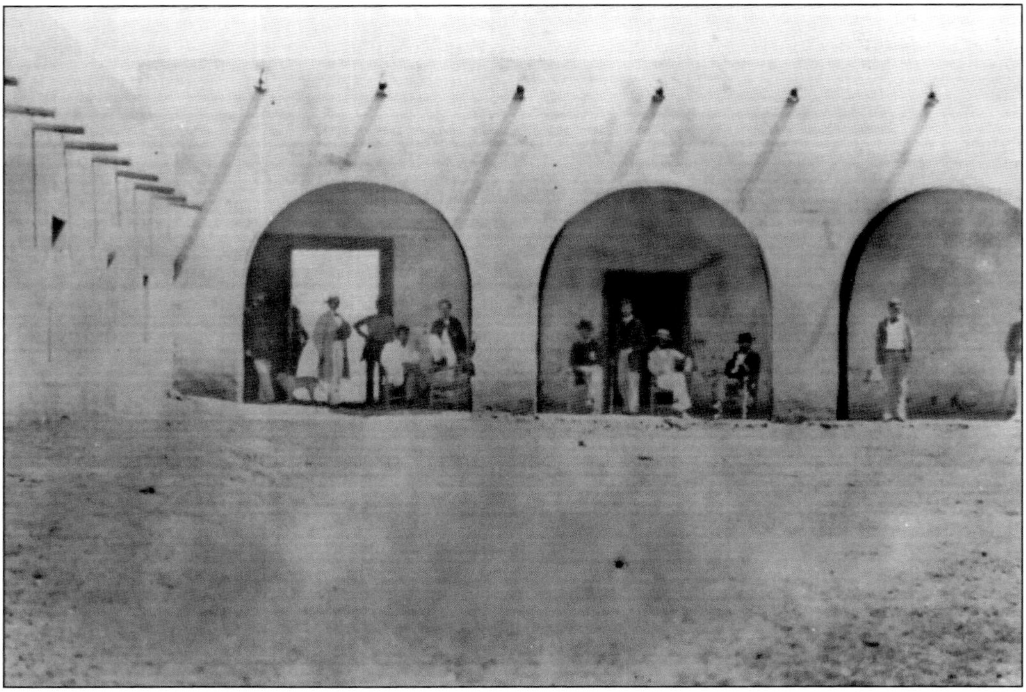

This Magoffinsville scene shows a gathering of men relaxing in the shade of a breezeway. This grouping of men has some in uniform, some in civilian clothing. This may have been before or after a midday meal.

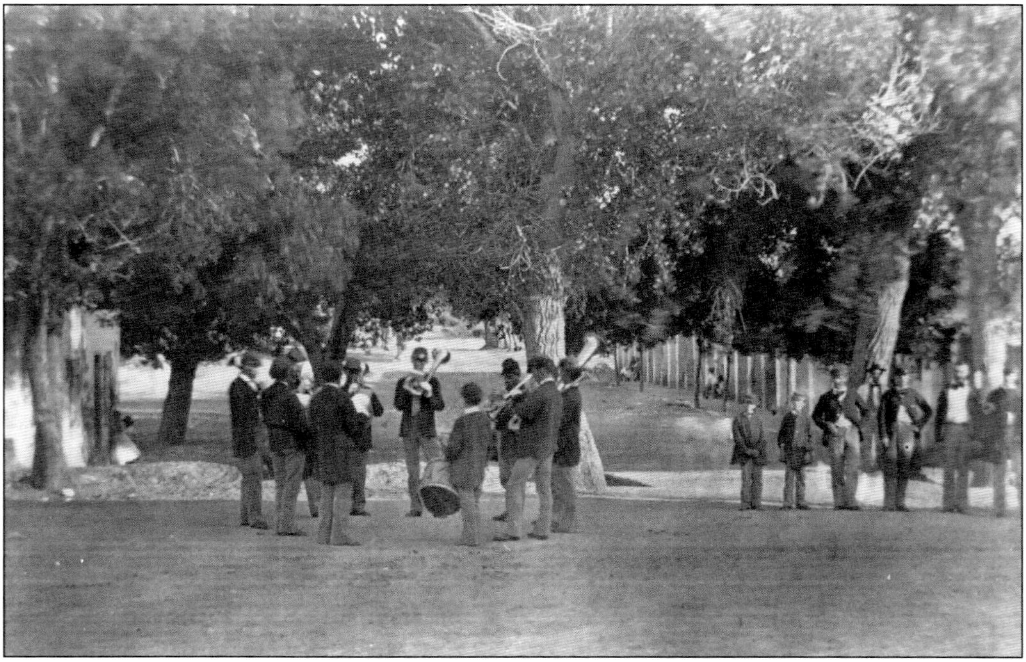

A regimental band is playing on South El Paso Street in the 1860s. There are young boys nearby wearing uniforms, likely the children of officers standing close by. Musicians were authorized in regimental units to blow bugle calls marking the next event during the day. They also played concerts for units and townsfolk when time permitted.

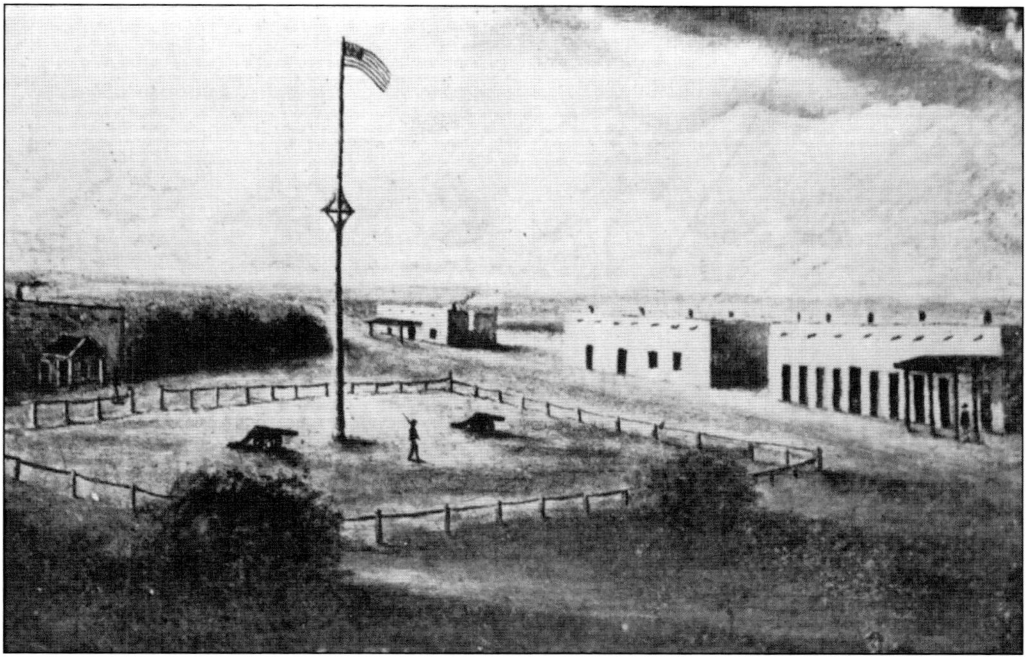

When the Magoffinsville post washed away in 1868, the Army moved three miles east and established Camp Concordia on Hugh Stevenson's ranch. There were no water sources nearby, so it had to be brought from the river three miles away. This was a barren post, not suited to planting a garden. But it was situated so the river would not wash it away.

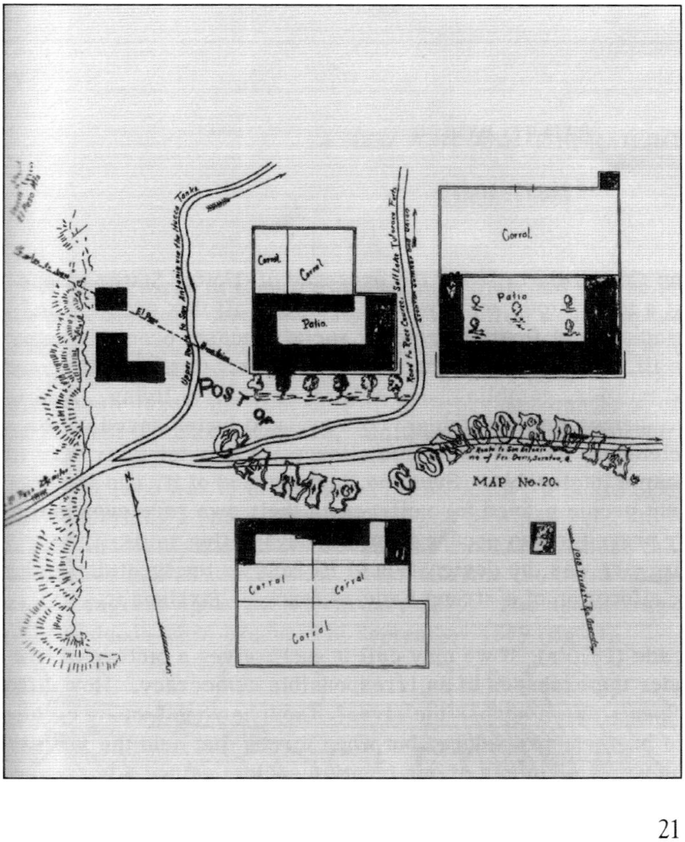

Camp Concordia was quickly renamed Fort Bliss. The post occupied about 10 acres. There were three large adobe buildings and several small ones. The buildings were one-story high with earthen floors and roofs. Although some distance from the Rio Grande, troops still suffered from dysentery and malaria.

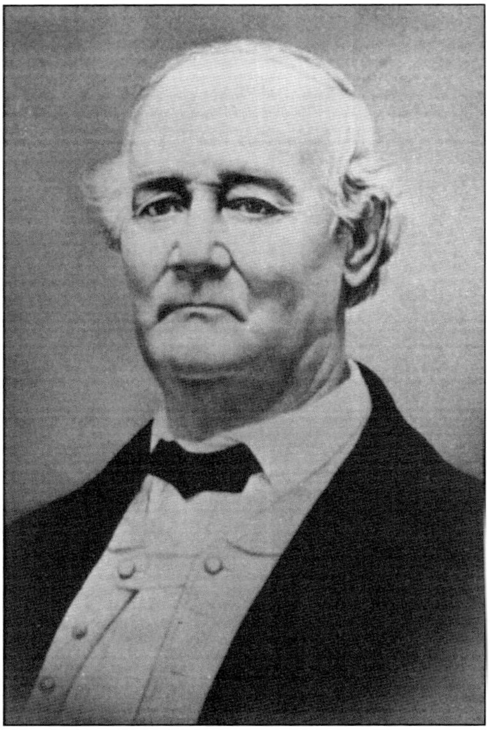

Rancher Hugh Stevenson rented the land and buildings to the Army, as had James Magoffin. This post was considered adequate but was still very uncomfortable. These quarters could house up to 200 soldiers; it was a small post like many other frontier posts. From 1870 to 1876, one or two companies of the African American 24th and 25th Infantry Regiments were there. The post was closed in January 1877.

The civil uprising in San Elizario known as the Salt War caused the Army to reestablish Fort Bliss in 1878. With the Concordia post being in ruins, troops were billeted in empty buildings around Pioneer Plaza (now San Jacinto Square) in downtown El Paso. This is a picture of the square in 1893, looking northwest. Soon the Army would decide to build a more permanent post.

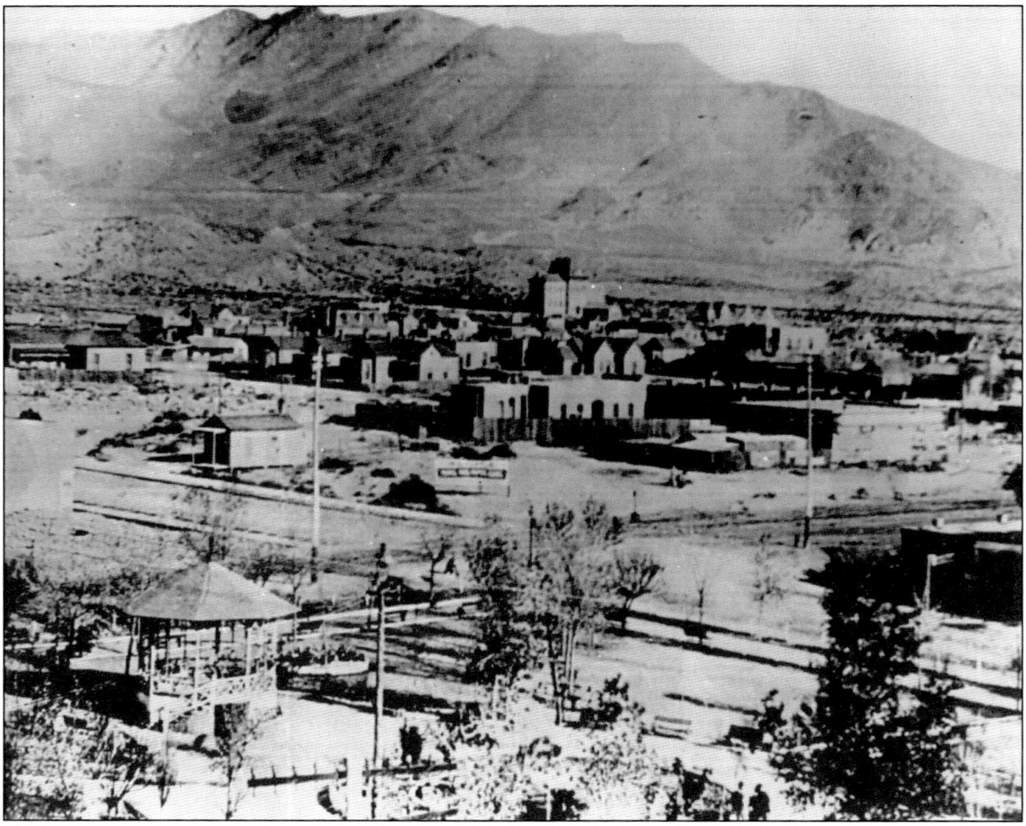

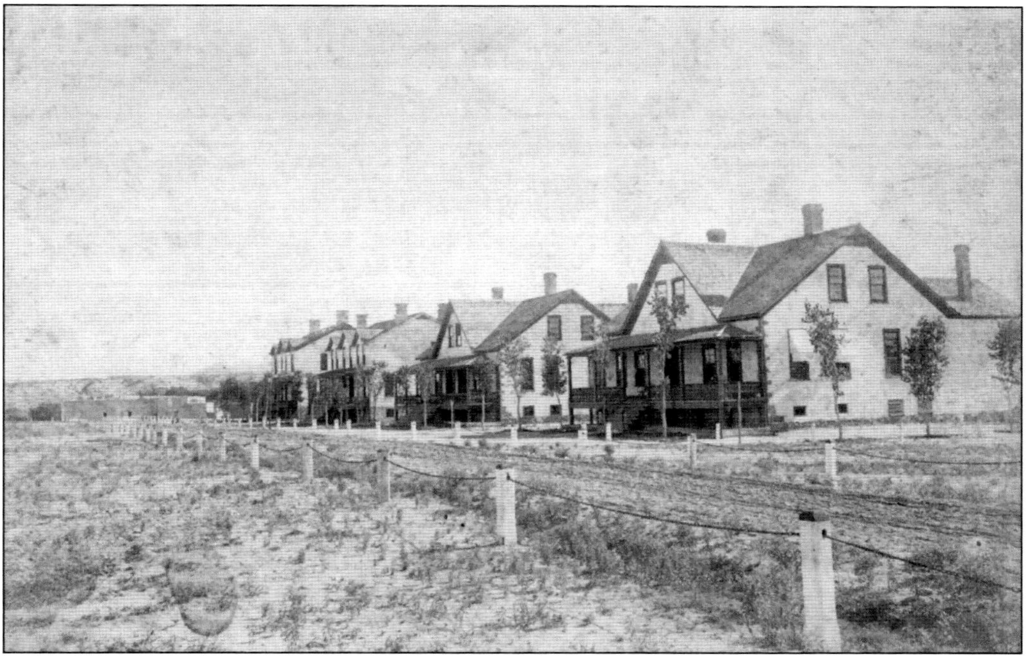

Rising Indian troubles once again called for a more permanent post. The Army bought about 135 acres of land west of downtown, near the old Simeon Hart gristmill. Congress provided $40,000 for the post, with the assumption that the troops would build it. This image shows Fort Bliss at Hart's Mill in the 1880s. These are officers' quarters, with a storage building in the distance.

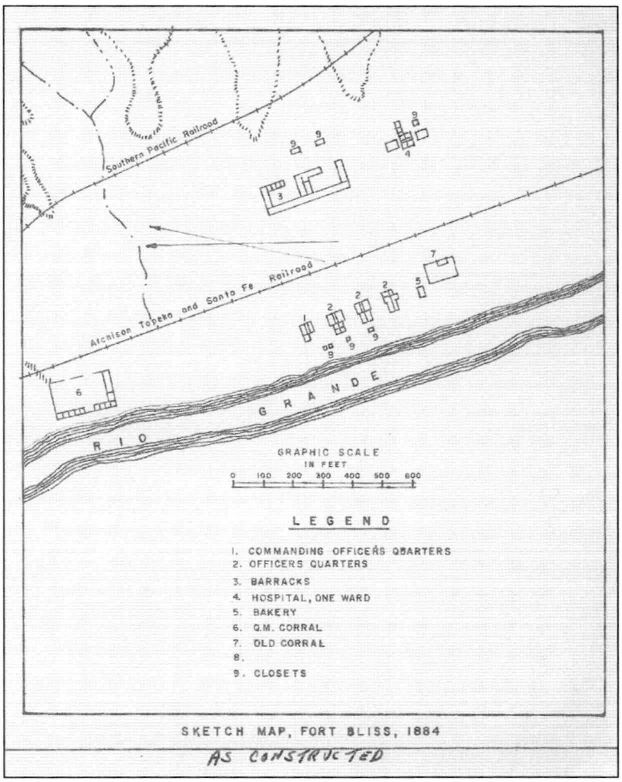

This 1884 map shows Fort Bliss at Hart's Mill. At the top are barracks, and below are the officers' quarters. The "closets" refer to outdoor privies. For the first time, the post had a structure that was truly a hospital, rather than buildings that served other purposes but were labeled thus. The map shows the tracks of two railroad rights-of-way through and adjacent to the post.

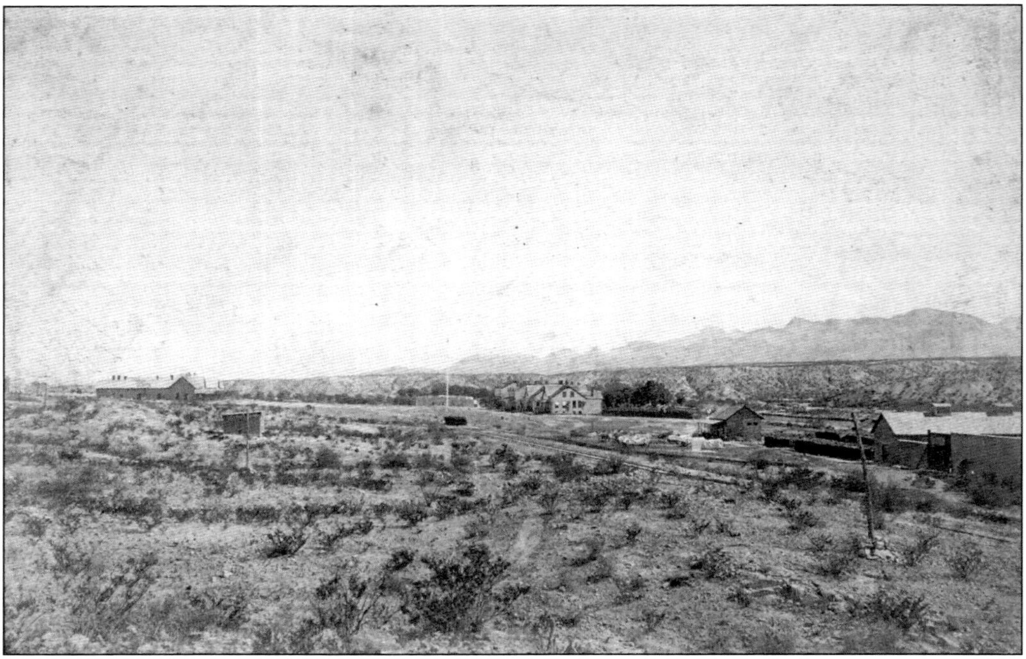

This is the Hart's Mill post in the afternoon, looking east. The quartermaster corral is in the right foreground, the officers' quarters are to the right, and the barracks are on the left. The railroad tracks are in the center and run past the flagpole. Today, this area is substantially built up on both sides of the river, with a major highway being constructed at this writing.

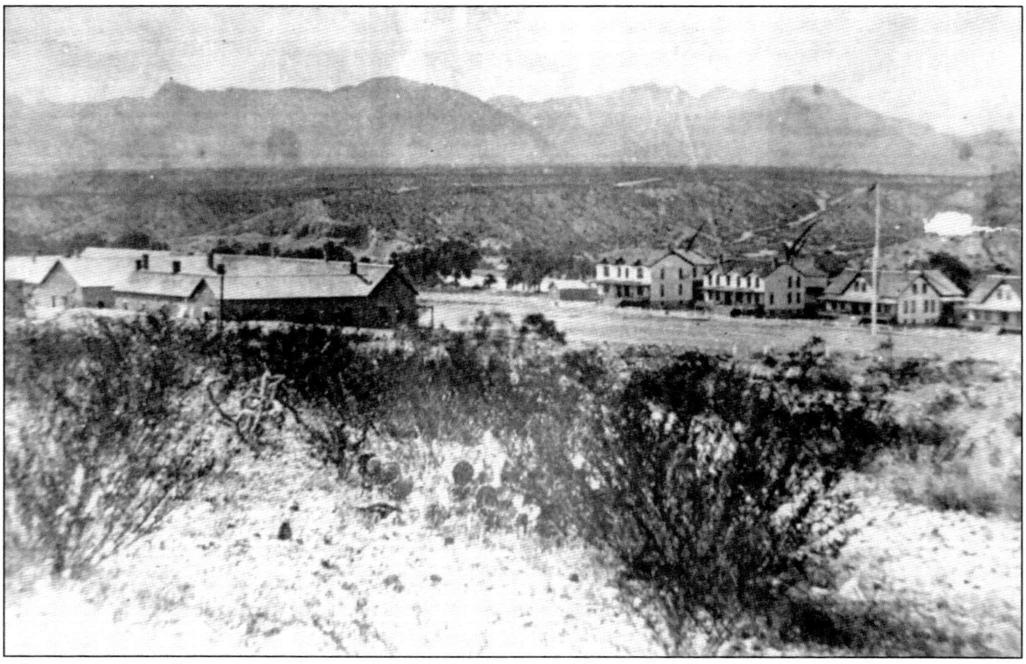

These are the officers' quarters in the early morning. The railroad tracks are clearly in evidence. The hills in the background are in Mexico, just across the Rio Grande. They are now covered with Mexican homes. In the distance are the Juárez Mountains. Only the two quarters on the left stand today on Paisano Drive.

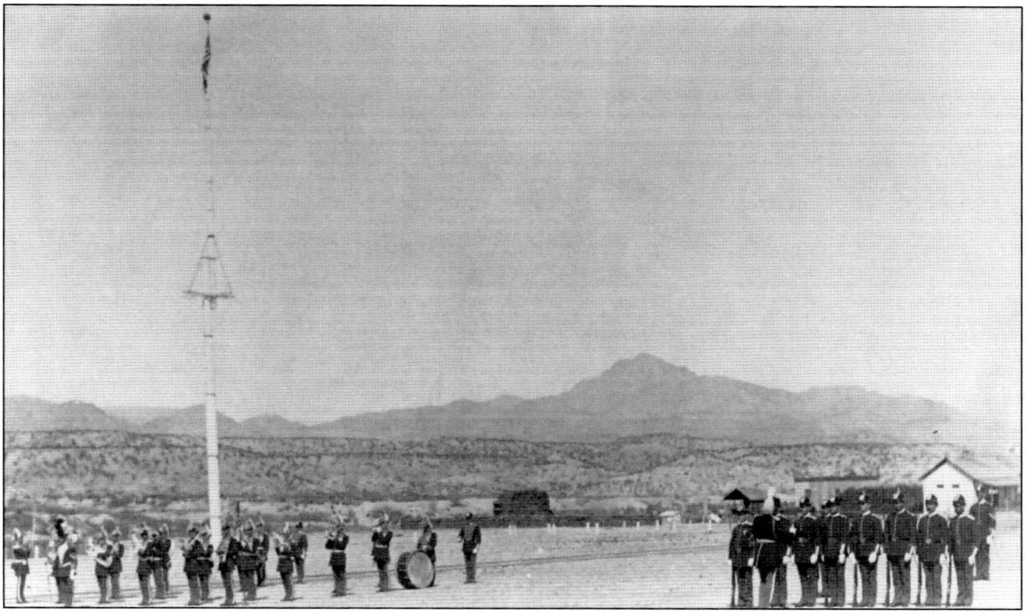

The regimental band and infantry troops in dress uniforms stand formal guard mount during retreat ceremony at day's end. Many of the troops in the 1880s were in the field pursuing Apaches as the Indian Wars were coming to an end. The Franklin Mountains are in the background. Today, two railroads and the interstate highway pass through here, with the University of Texas El Paso behind them.

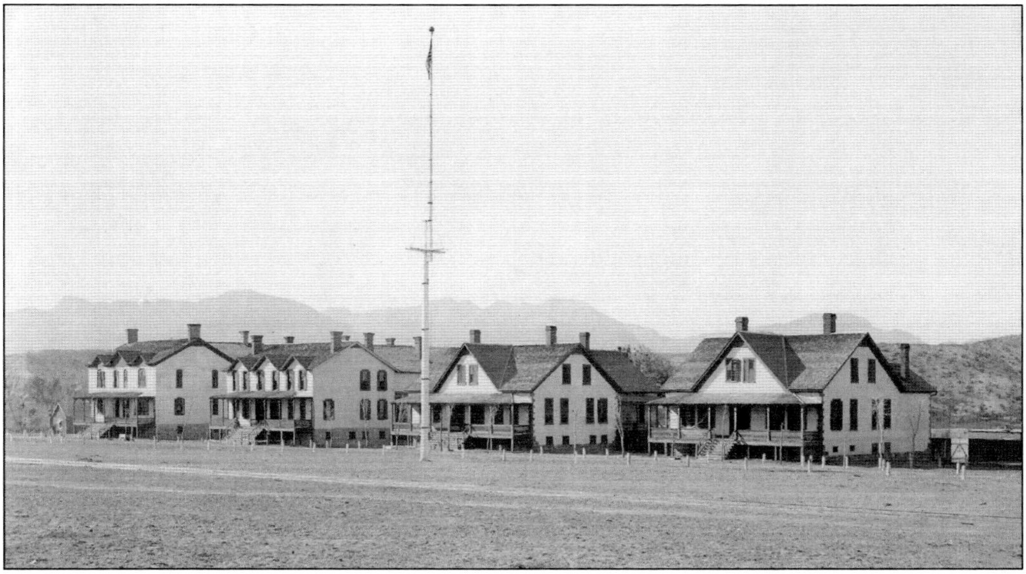

The Hart's Mill officers' quarters stand side by side in the early morning. This is looking south, with the Rio Grande just behind the quarters and the hillocks of Mexico just beyond. The Juárez Mountains are in the distance. Today, the hills beyond these houses are covered with Mexican houses, and a border fence barrier separates the quarters from the river.

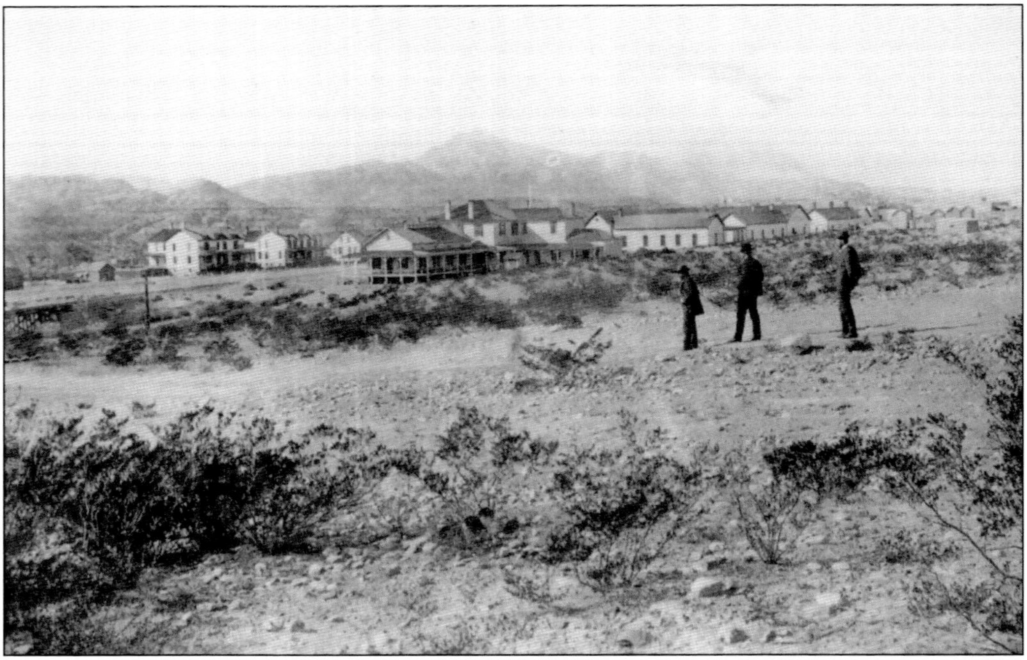

This is the post in 1890, looking west. The Army was preparing to sell the land and buildings to move the post, Congress having authorized funds to construct a new Fort Bliss. The hospital is in the foreground, and the officers' quarters are at left. The post was to be sold through public and private sale. The Apache conflicts ended in 1886, and a new era was beginning.

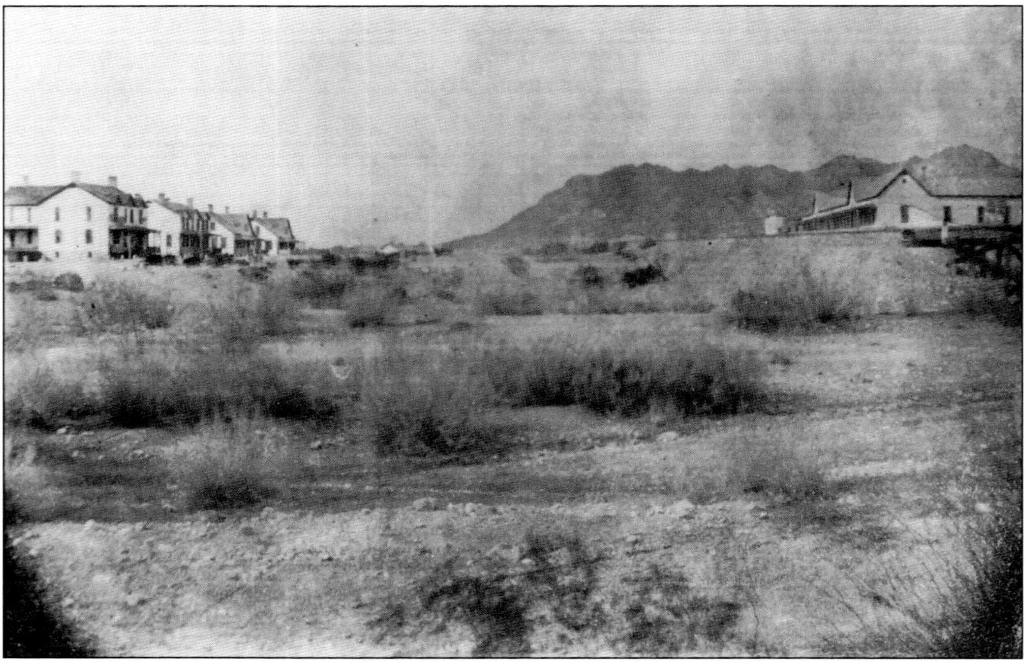

This is a last look at Fort Bliss at Hart's Mill, with Mount Cristo Rey in the distance. Unseen is the Rio Grande to the left of the officers' quarters. A railroad trestle for the Atchison, Topeka & Santa Fe Railroad is at right. The appearance of trains during the day and night had to be disruptive, so it was time for Fort Bliss to move once again.

Two

Brick and Infantry

The Post on La Noria Mesa

In 1890, the US Census Bureau declared the American frontier closed. Railroads linked major cities and towns from coast to coast. No less than five railroads converged on the city of El Paso, incorporated in 1873. With many far-flung small posts throughout the West, the Army concluded that it was time to close and consolidate these penny-packet posts, a formal policy that is recognized today as Base Realignment and Closure.

Fort Bliss had survived two closings and five different moves during the Army's frontier period. After shifting the Fort Bliss post from five different locations, in 1890 the War Department made the decision to build a major Army installation in the El Paso area. Instead of the rented adobe and brick quarters that were built with soldier labor, this post was to be a modern installation with permanent, contract-built structures to house up to four infantry companies. The selection of a site for the new post was a cooperative effort between the Army and El Paso city fathers. With the city and railroad encroachment on the Fort Bliss at Hart's Mill, a new location that would allow for more troops and better training areas was imperative.

The Army selected a site on a place called La Noria Mesa, about five miles northeast of the city. The owners of the land were prominent El Paso citizens: banker Charles R. Morehead, Judge Bennett H. Davis, and Capt. Thomas J. Beall, former Confederate officer and El Paso lawyer. The three agreed to sell the land to local subscribers, who would then turn the land over to the Army. Local subscribers signed up and signed up again for an additional offering when the Army requested more land.

The intent was to build a modern military installation with up-to-date comforts for the troops. No more troop labor; this time the Army contracted with several firms to build the post from scratch. The Army dispatched one of its best quartermaster officers to oversee the project, and by 1892, a new, modern post existed where before there was nothing but desert.

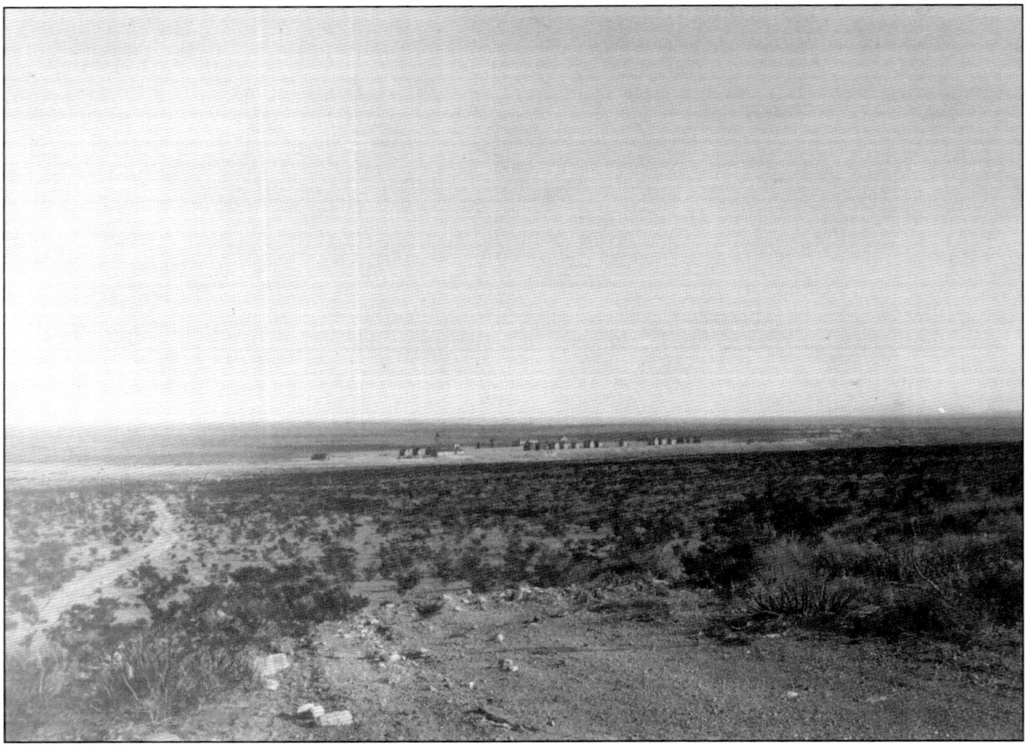

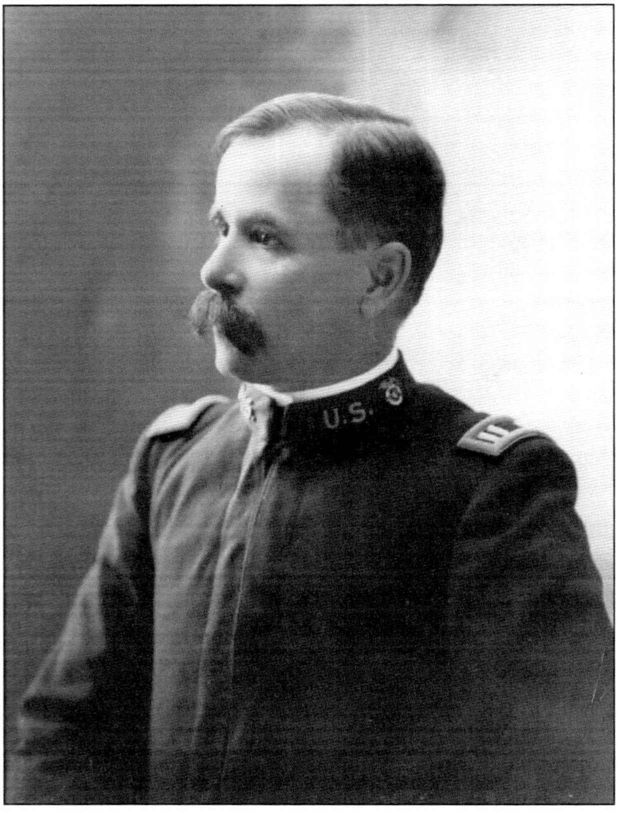

This is the newly built post on La Noria Mesa, taken from the Franklin Mountain high ground to the northwest. The post stands alone on the barren prairie, with only the two barracks buildings, the post headquarters, the water tower and pumping station, and officers' quarters. This was the original post, isolated from the town but laid out as a modern military installation at the end of the 19th century.

Capt. George Ruhlen was an 1872 graduate of the US Military Academy. He was an experienced Quartermaster Corps officer who had overseen the construction of other Army posts. He arrived in El Paso on March 14, 1891, to survey the La Noria Mesa site and lay out the locations for the various post buildings. Captain Ruhlen and his assistant moved into the Sheldon Building downtown and went to work.

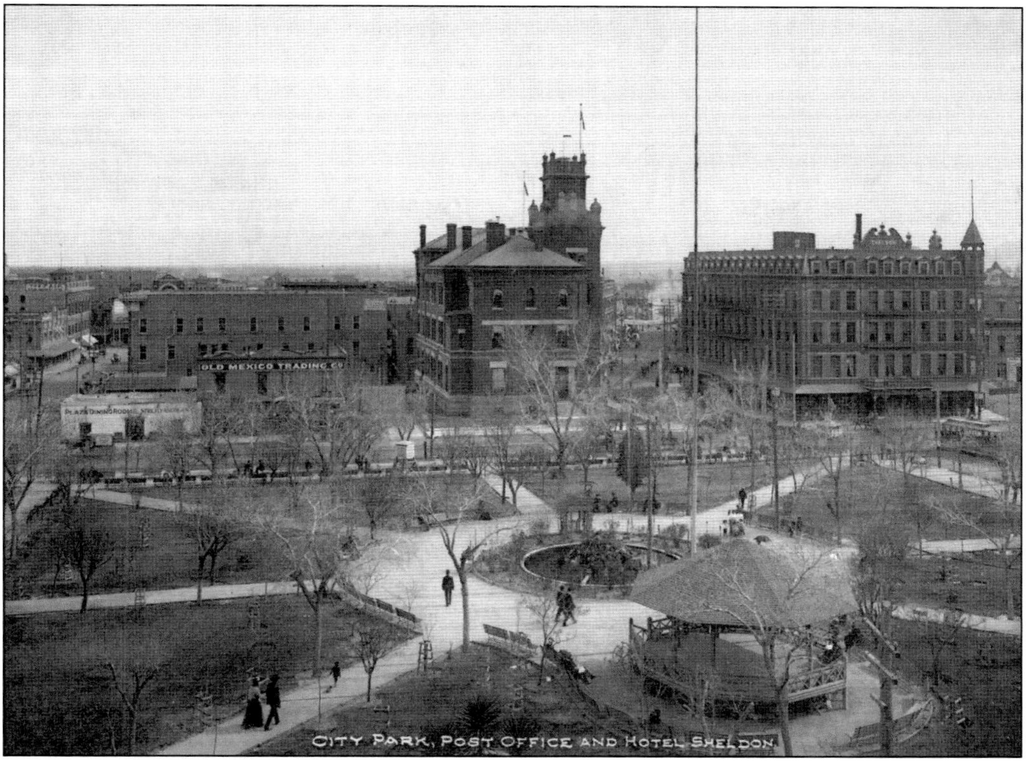

Captain Ruhlen and his party occupied space on the top floor of the Sheldon Hotel on San Jacinto Plaza. Ruhlen was qualified as both an architect and engineer. He had his family with him, expecting the project to take several years. Ruhlen was described as a man of medium height, compact build, and athletic appearance. He was well received by the city fathers of El Paso.

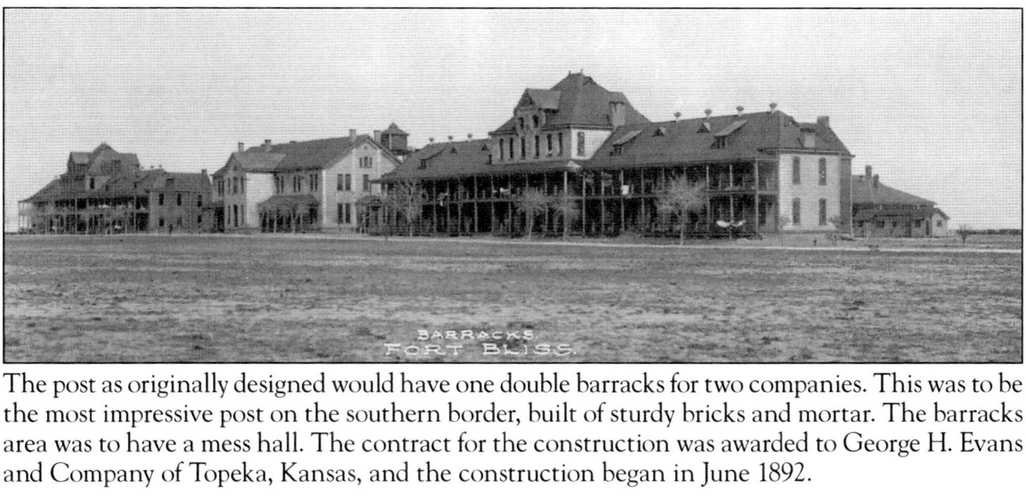

The post as originally designed would have one double barracks for two companies. This was to be the most impressive post on the southern border, built of sturdy bricks and mortar. The barracks area was to have a mess hall. The contract for the construction was awarded to George H. Evans and Company of Topeka, Kansas, and the construction began in June 1892.

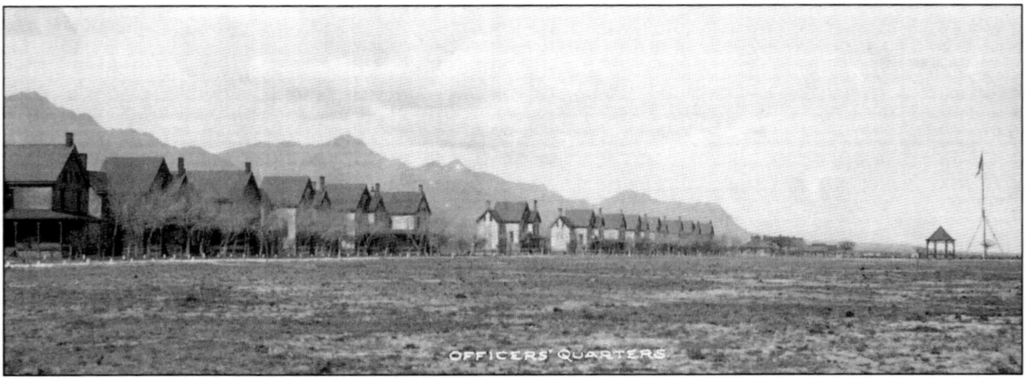

Opposite the barracks area were to be four sets of captains' quarters and six sets of lieutenants' quarters. By November 1892, construction was well under way. Walls were up, water lines were going in, and the buildings were otherwise taking shape. There were problems with the quality of bricks, but this was solved by changing brick contractors. Today, some houses are of yellow brick and some of sturdier red brick.

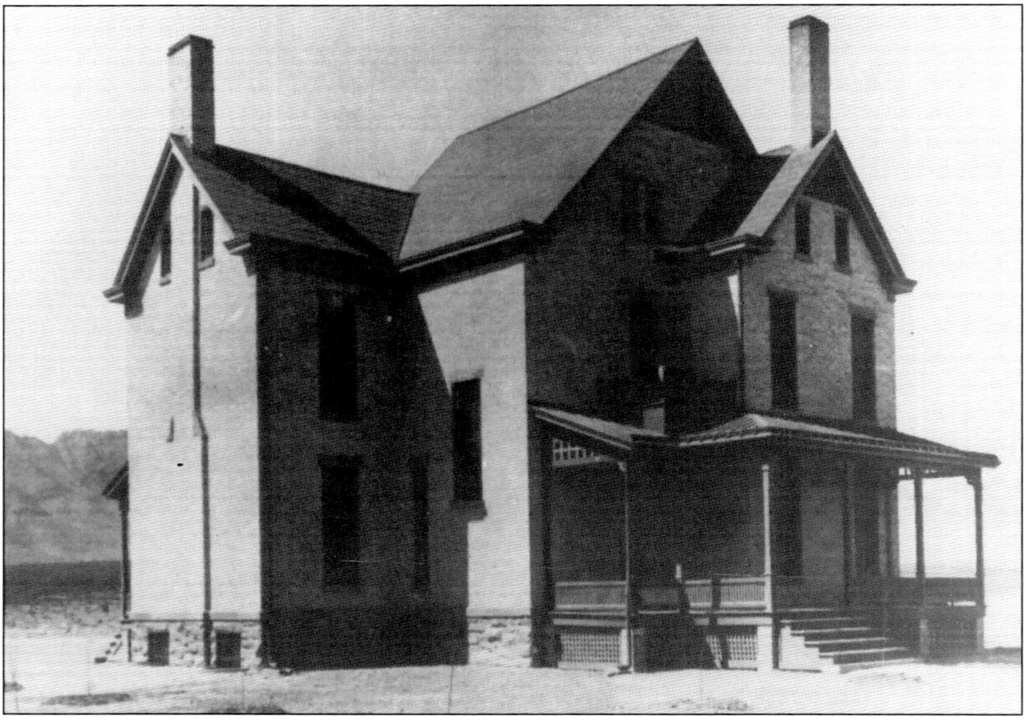

The lieutenants' quarters were to be two-story buildings following a current Army quartermaster design. Windows were tall, allowing for good ventilation of the quarters in the warmer months. The quarters had indoor plumbing for bathrooms. Originally, lighting would come from kerosene lamps. Electric lighting would not come until 1911. The quarters had broad porches and a sleeping porch over the veranda.

The captains' quarters were also two-story buildings, with wide verandas in front and back. The entryways had double doors with screens, and as with the lieutenants' quarters, windows were double-hung tall glass. Heating for both came from fireplaces. Both had substantial basements, excavated in the ground by blasting due to hard rock. The architecture was a modified Queen Anne style popular in the 1890s.

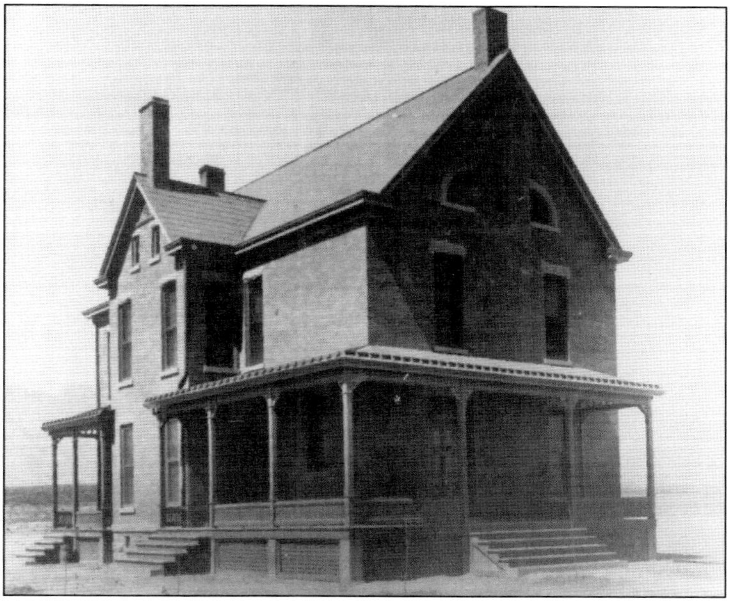

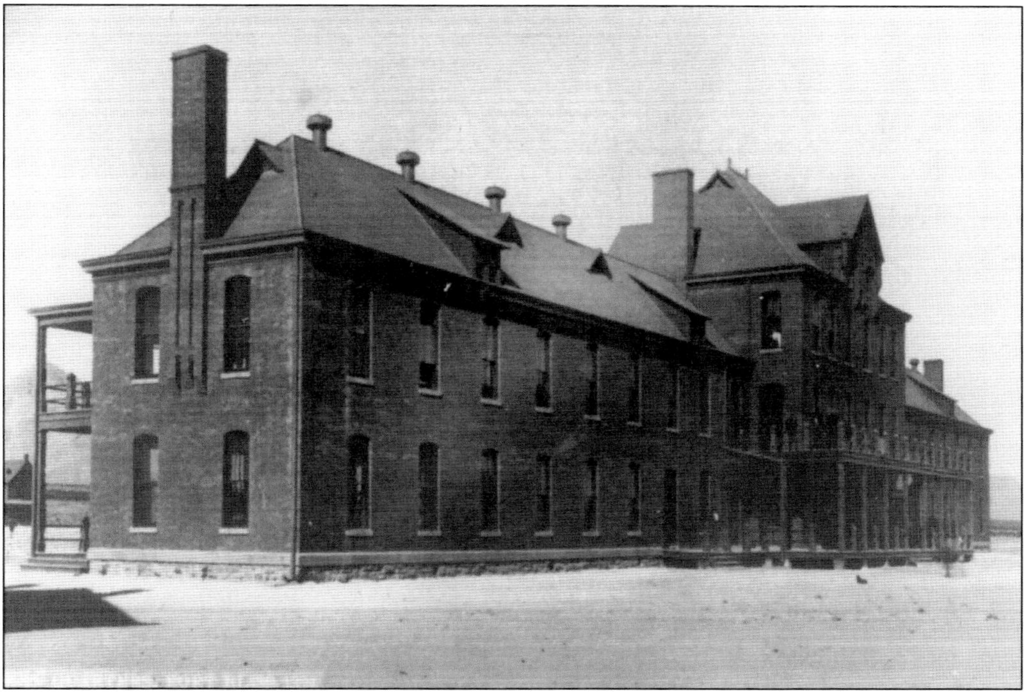

Now Building 111, this two-company barracks was completed in 1893. This was an addition to the original plan for one-company barracks. In August 1892, Captain Ruhlen received orders to expand the plan to another company barracks and six more officers' quarters. Two lieutenants' quarters were deleted, but that still added four officers' quarters to the plan. These enlisted quarters were substantial, providing greater comfort to the troops.

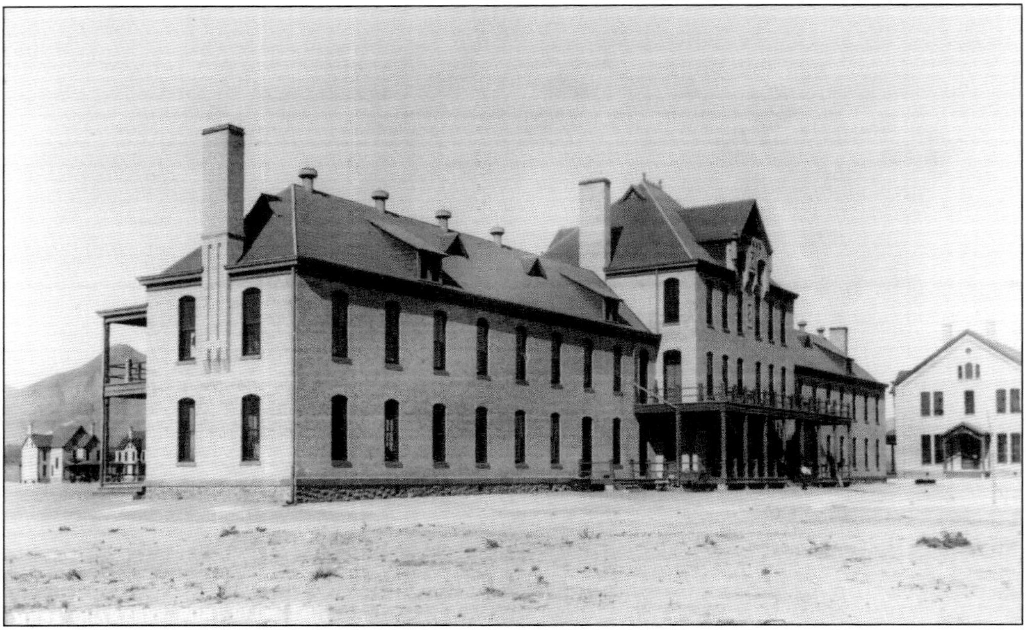

Now Building 13, this is the matching two-company barracks to Building 111. Both were standard quartermaster building designs for the 1890s. Heating came from fireplaces at the end and center of the buildings. The barracks had open bays, with no privacy for the troops who lived there. Bunks were lined against the walls. Cooling in summer came from opening the windows.

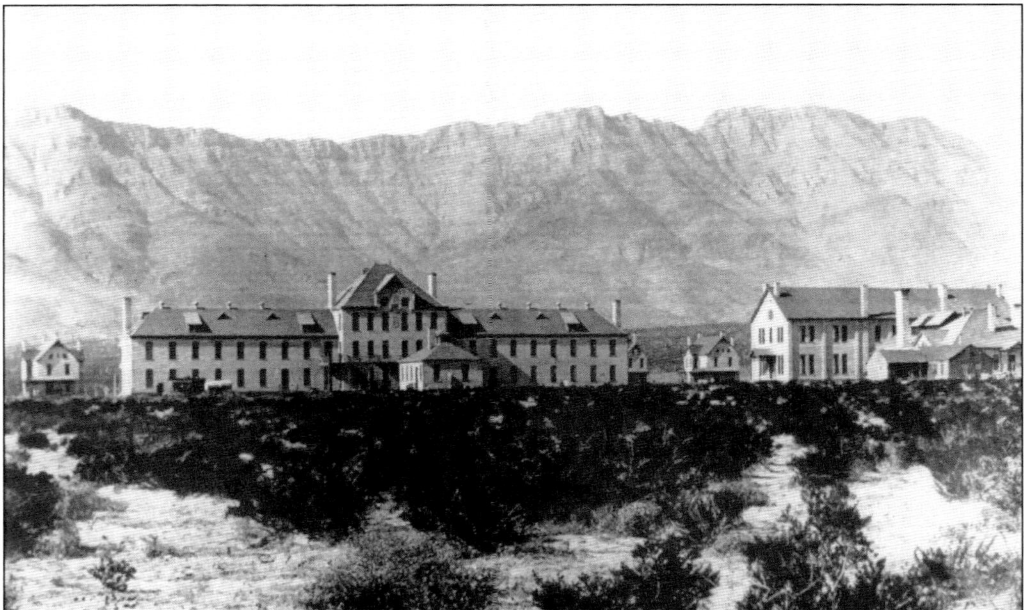

This rear shot of the other company barracks and the headquarters building shows some of the officers' quarters beyond and the prominent Franklin Mountains in the distance. In the foreground are clumps of creosote bush, cactus, scrub mesquite, and ocotillo. Despite the modern construction, the surroundings were still those of a frontier post. But it would not be very long before the city expanded out to the post.

This rear view of Building 113 shows the extensive two-story barracks, the bathhouse and latrine, and most importantly the water tank and steam plant. It included a complete water supply for the entire post and a sewer system. The well was 230 feet deep, and the steam pump could produce 600 gallons of water per hour.

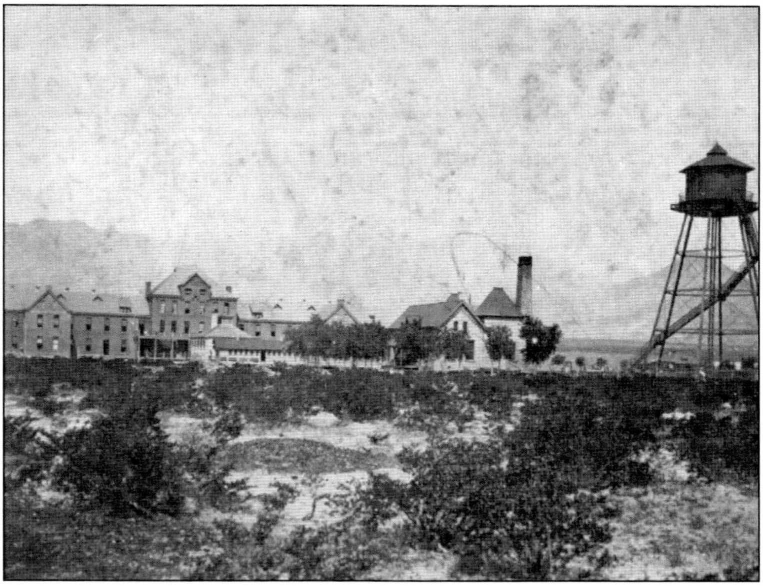

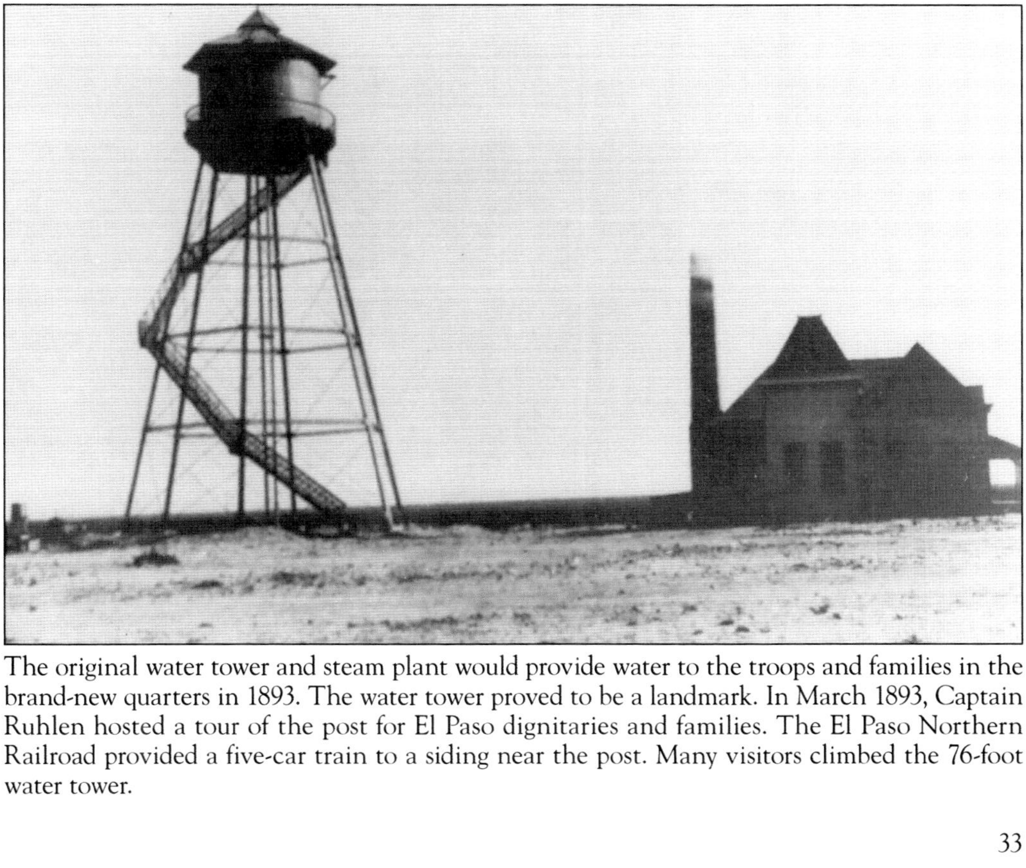

The original water tower and steam plant would provide water to the troops and families in the brand-new quarters in 1893. The water tower proved to be a landmark. In March 1893, Captain Ruhlen hosted a tour of the post for El Paso dignitaries and families. The El Paso Northern Railroad provided a five-car train to a siding near the post. Many visitors climbed the 76-foot water tower.

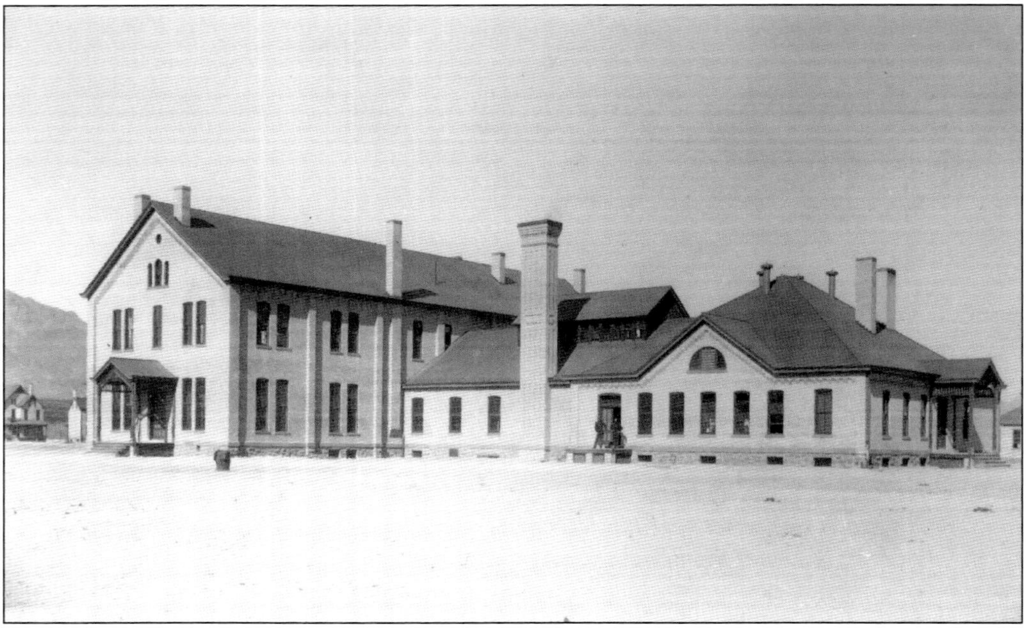

The headquarters building was an important part of the new post. Four companies of troops required a regimental headquarters. The new headquarters provided space for administration as the post construction proceeded. As construction progressed, a unit of the 23rd Infantry moved from the Hart's Mill post to provide an interior guard on the post. This unit reported to the commander and staff in the headquarters.

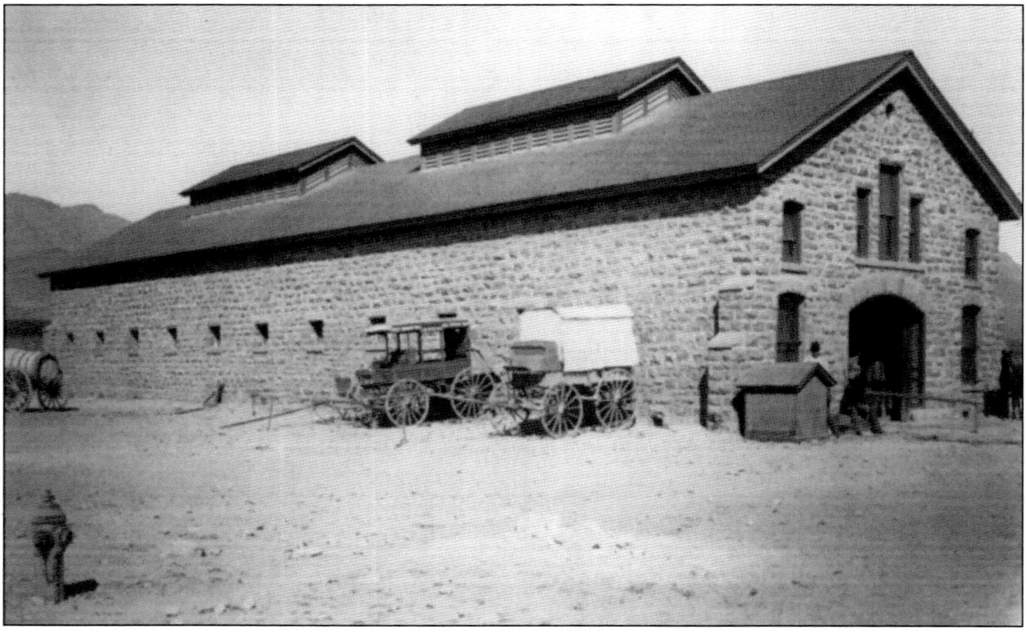

The commissary was built of mortar and locally available rubble rock. This served as the issue point for foodstuffs, including flour, sugar, beans, coffee beans, vinegar, and other staples that served as troop rations. This building subsequently served as a laundromat, thrift store, and meeting place for post organizations throughout the years. Built of quarried and dressed stone, it is a substantial structure.

In September 1893, the 18th Infantry Regiment was ordered to relocate to the new Fort Bliss from Fort Clark, Texas. The regimental staff, four companies, and the band were to arrive on October 1. Commanding them would be Col. Henry Lazelle, the same officer who was wounded at Dog Canyon 34 years before. Lazelle was on sick leave at the time, so the first commander was Capt. William McLaughlin.

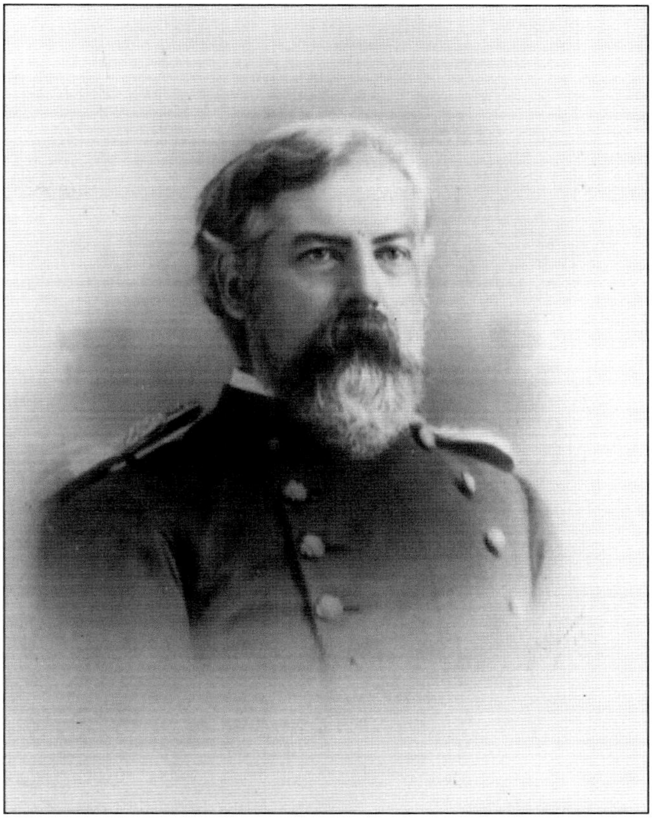

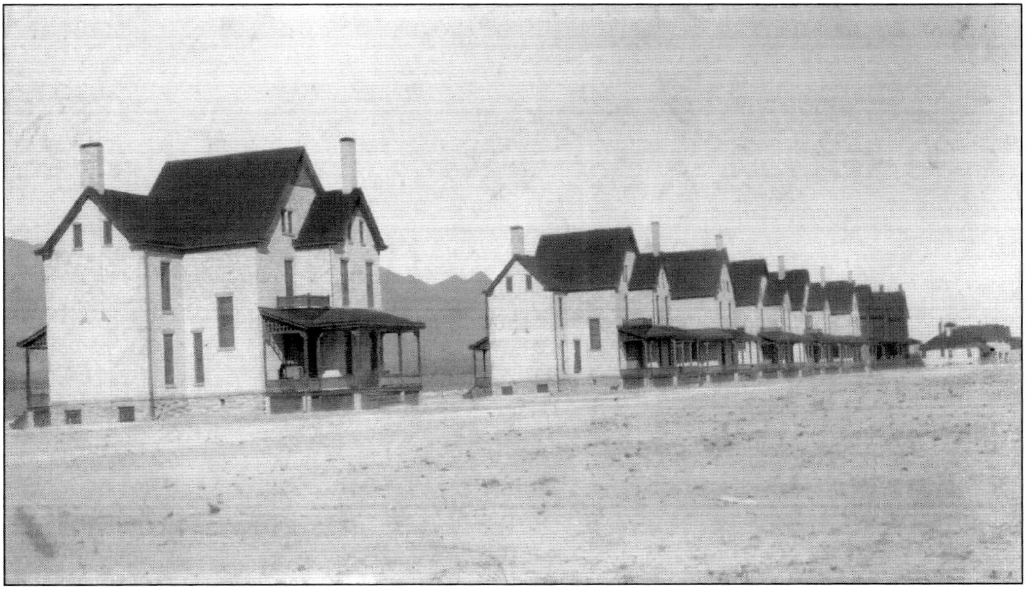

The officers of the 18th Infantry Regiment moved into the excellent quarters. Six sets of quarters were built of yellow brick and the seventh of more durable red brick. The officers who occupied these quarters were, from left to right, Lieutenant Evans, 24th Infantry; Lieutenant Ano, 18th Infantry quartermaster; Captain Babs; Captain Hood; Lieutenants Greg and Goole; Captain Minick; Lieutenant Beall, 18th Infantry adjutant; and Captain Hinton.

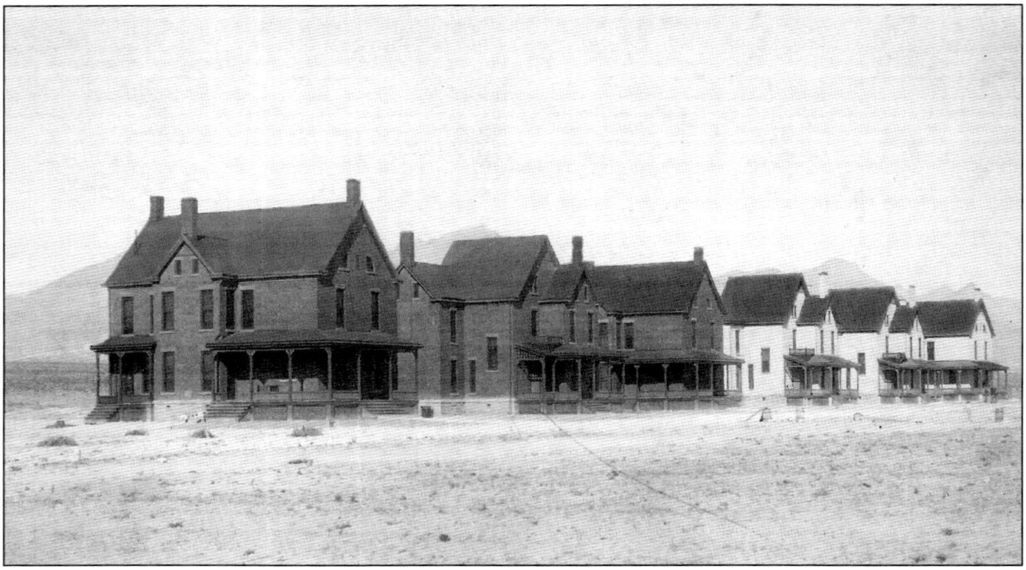

The names listed on the back of this picture are Dr. Tailor; Dr. Chapin; Captain McLaughlin, commanding the regiment; Lieutenants Moore, Whitmont, and Baldwin; Lieutenant Dural; and Colonel Parker. Lieutenants without families shared quarters. McLaughlin commanded until Colonel Lazelle arrived from sick leave in 1893. However, Colonel Lazelle did not stay long. He departed again on sick leave in 1894, and Col. Daingerfield Parker replaced him.

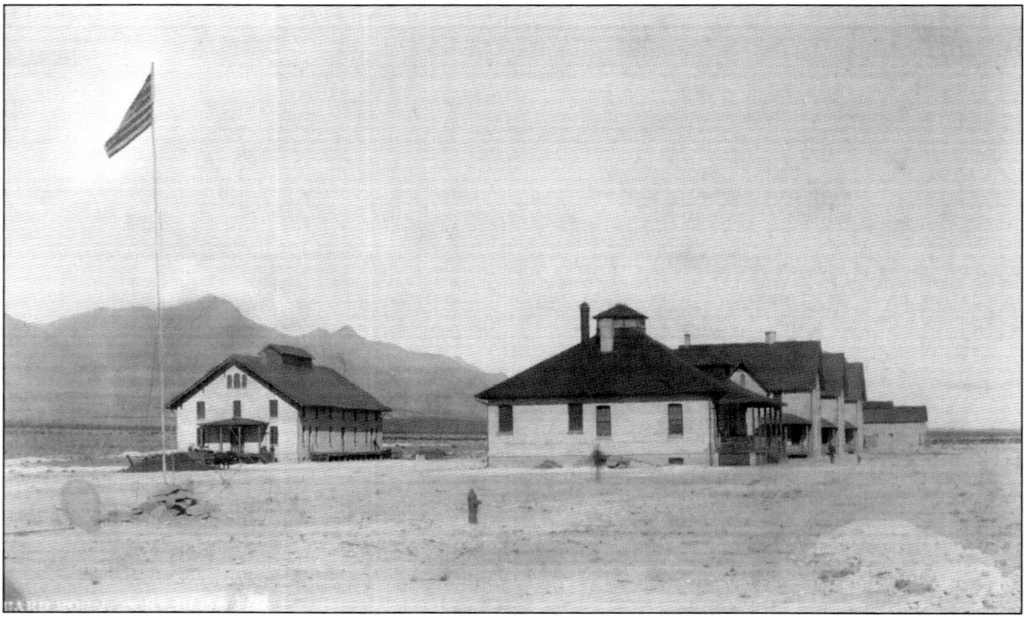

The first flag at Fort Bliss on La Noria Mesa was raised by Janet Hood, Captain Hood's wife. This image shows the flagpole, the commissary and quartermaster storehouse, and the guardhouse. On Friday, April 5, 1894, a terrible storm of sand and wind raged all day. The barracks housing H Company and two noncommissioned officer quarters were badly damaged. Troops moved to Fort Sam Houston until the buildings could be repaired.

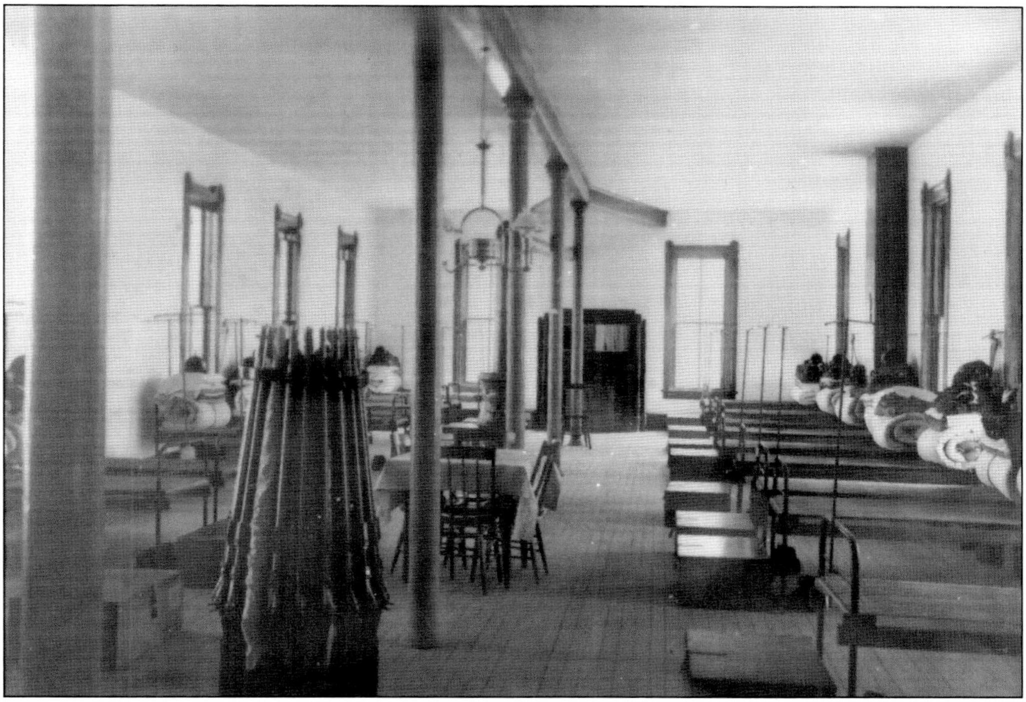

This was a barracks room around 1894. Lighting came from kerosene lamps, with electricity not coming until 1911. Bunks were "stacked" for the day to keep dust off. Weapons racks held Springfield trap-door black powder rifles firing .45/70 cartridges. There were footlockers and hanging racks at the ends of the bunks for soldier belongings. Life was spartan in Army barracks in the 1890s.

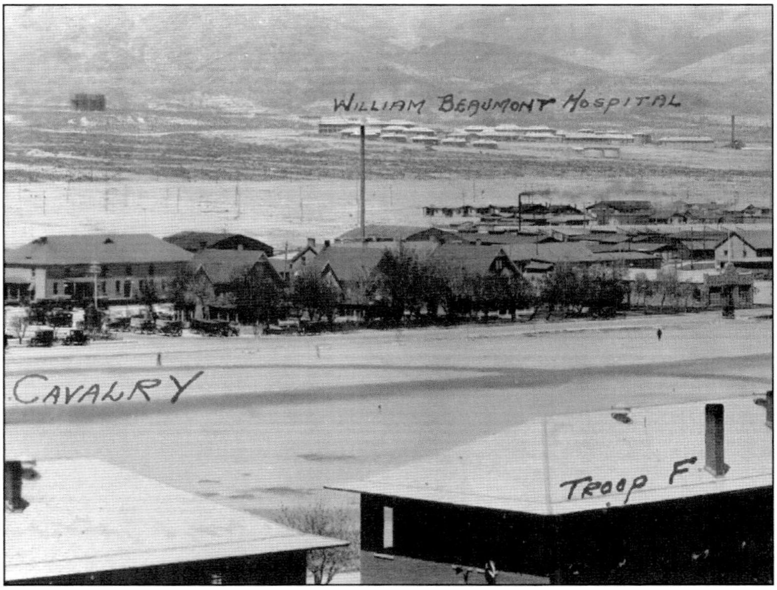

The three buildings in the center of the picture are staff NCO quarters. They were fourplexes, capable of housing four families each. These were in the original plans for Fort Bliss and provided some of the best housing for sergeants anywhere in the Army. These quarters stood until they were torn down in 1966.

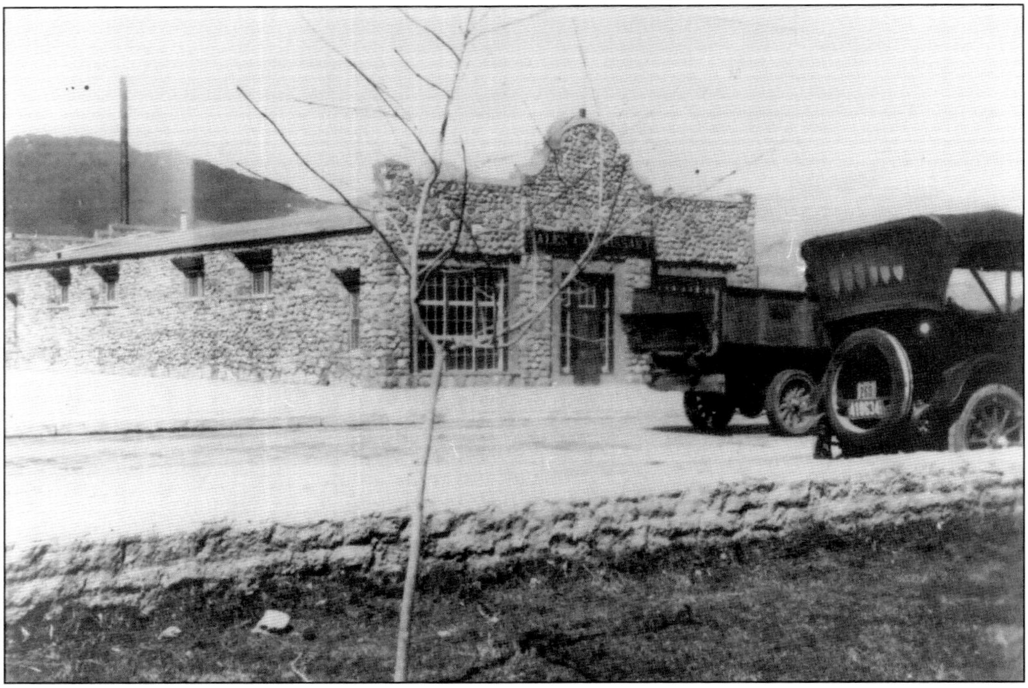

This was the original 1893 commissary sales office. The picture was taken in the 1930s, after the structure had been substantially rebuilt. In the 1930s, this structure was used as a signal equipment workshop and storage building. It still stands on Fort Bliss today. By the 1890s, refrigeration was beginning to come into use, but Fort Bliss would not have real refrigeration until the post was electrified in 1911.

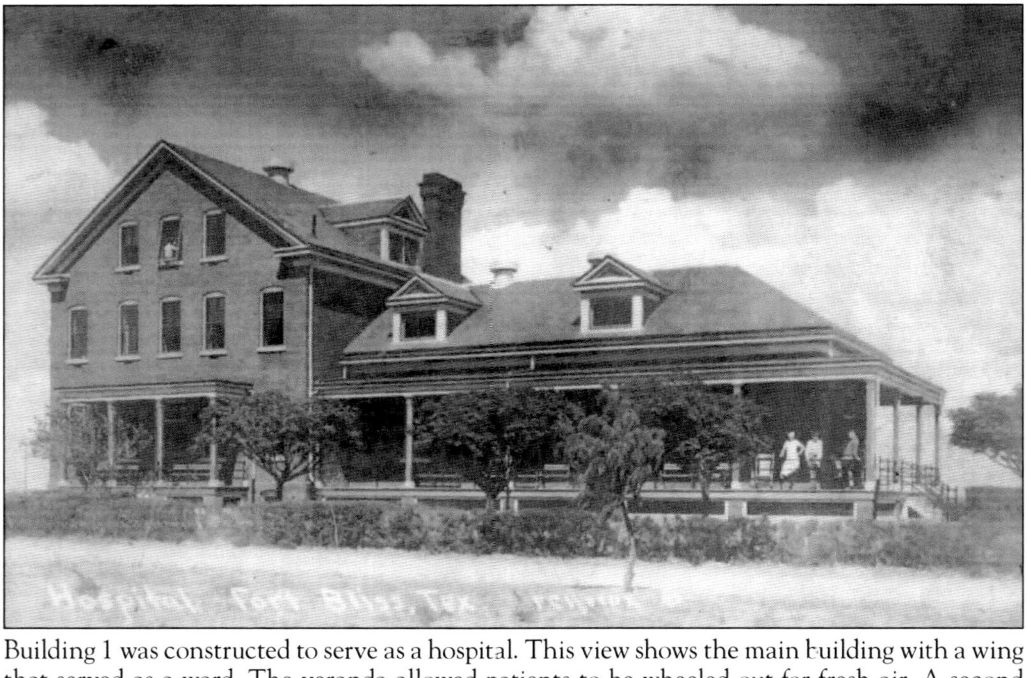

Building 1 was constructed to serve as a hospital. This view shows the main building with a wing that served as a ward. The veranda allowed patients to be wheeled out for fresh air. A second wing at left was added later. Medical stewards and staff stand on the porch.

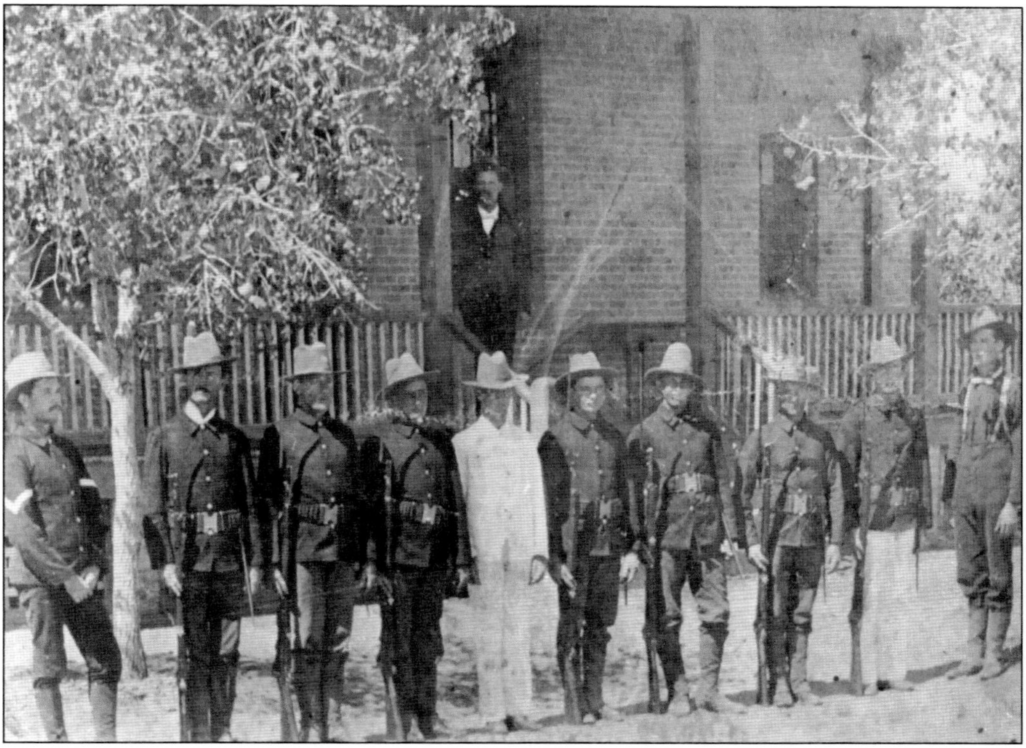

Troops conduct a formal guard mount in the 1890s. Soldiers were armed with the 1873 Springfield trap-door rifles. These were single-shot rifles that fired the formidable .45/70 black powder cartridge. Troops are wearing the 1892 sack coats with light blue trousers. The hats are 1889 campaign hats. They also wear the Mills web cartridge belts. The soldier at center wears a white fatigue uniform, and one soldier wears white trousers.

This is another picture of a guard mount around 1898. These troops are cavalry soldiers, probably 5th Cavalry. They replaced some of the 18th Infantry that moved to Fort Sam Houston in 1895. This was a dreary day in front of the officers' quarters and shows how bleak the post was at that time. Troopers are wearing the blue sack coat and are carrying 1873 single-shot carbines.

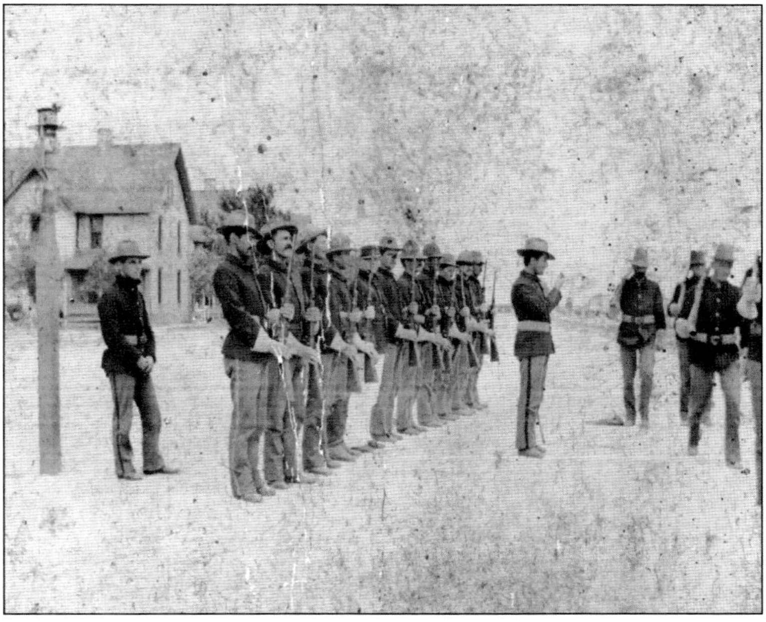

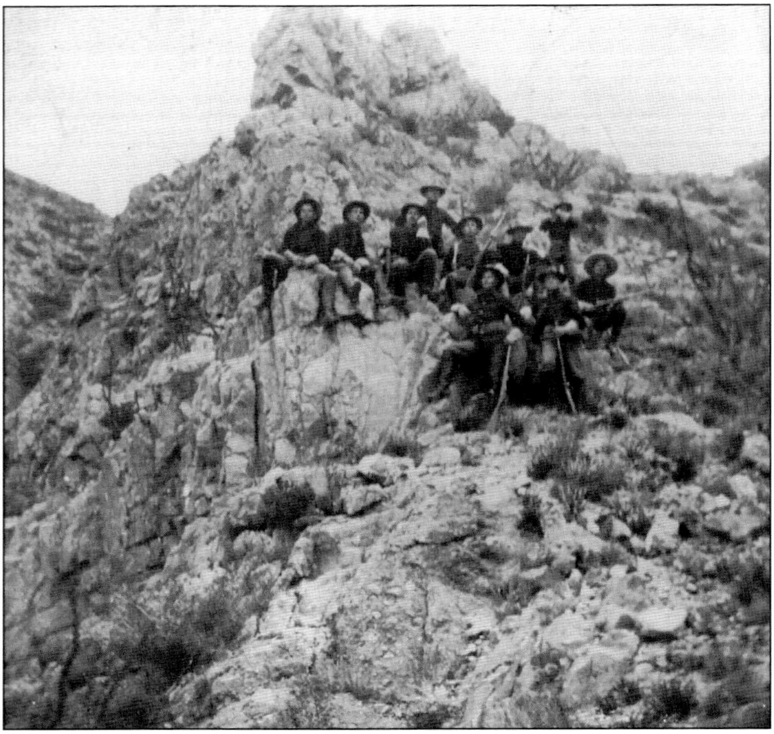

A squad of F Troop, Texas Volunteer Cavalry climbs Mount Franklin, west of the post. Troopers wear high cavalry boots, light blue trousers, dark blue cavalry shirts, and Mills cartridge belts. They are carrying the 1873 .45/55 trap-door carbines. All the regulars had gone to serve in the Spanish-American War. These troopers and troops of the 3rd Texas Infantry served at the post until the regulars returned in 1899.

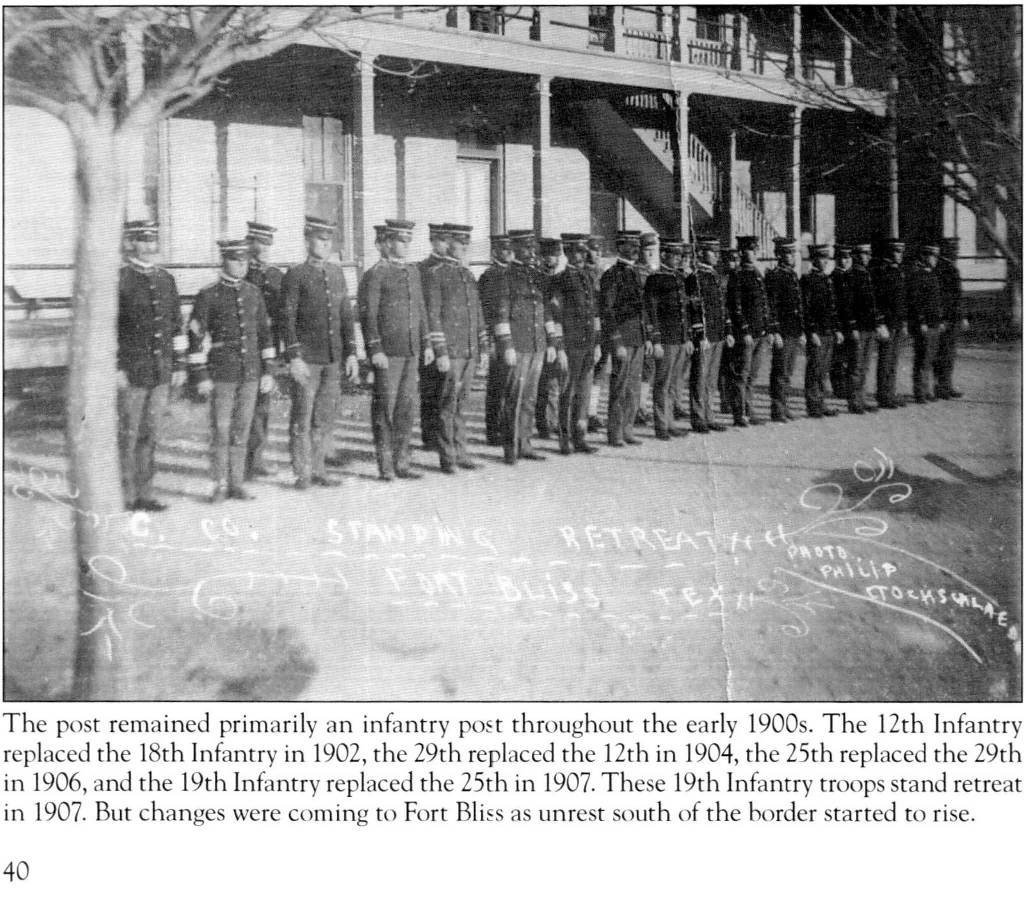

The post remained primarily an infantry post throughout the early 1900s. The 12th Infantry replaced the 18th Infantry in 1902, the 29th replaced the 12th in 1904, the 25th replaced the 29th in 1906, and the 19th Infantry replaced the 25th in 1907. These 19th Infantry troops stand retreat in 1907. But changes were coming to Fort Bliss as unrest south of the border started to rise.

Three

Leather and Revolution

Fort Bliss and the Mexican Revolution

The larger post on La Noria Mesa made it possible to concentrate troop units and permanently base troops near the Mexican border. The railroad allowed for rapid movement of forces and supplies. By 1893, infantry troops were comfortably established in their new home on the mesa. Troops of the 18th Infantry Regiment had moved from Fort Clark, Texas. The post had a population of over 500, the largest any Fort Bliss post had ever had. Volunteers replaced regulars during the Spanish-American War, but regulars rapidly returned to the border post when the war ended. In 1910, one of the great social upheavals of the 20th century began. This was the great Mexican Revolution. This conflict was on the doorstep of the United States, and the revolution was to last 10 years. This placed Fort Bliss in a strategic position to support protecting the border from incursions and to control operations to patrol the border.

The revolution's conflicts were primarily in central and northern Mexico, with many battles occurring near to the international border. In May 1911, Generals Francisco "Pancho" Villa and Pascual Orozco attacked Cíudad Juárez and successfully took the city. Many El Paso citizens went to the border to see the fighting, and there were US casualties and property damage on the American side. After this, Pres. William Howard Taft dispatched 11,000 cavalry troops to the Mexican border. In 1914, Pres. Woodrow Wilson ordered the movement of the 8th Infantry Brigade from the Presidio of San Francisco to Fort Bliss. As events unfolded, the post became a cavalry post, as infantry did not have the mobility to patrol the border areas.

The border region around El Paso became an active spectator and even participant in the revolution. Smuggling of arms and ammunition made some El Paso merchants rich. The post was the somewhat reluctant host of about 6,000 Mexican refugees in 1914. And in 1916, the post became the support base for the US Punitive Expedition into Mexico. All of this made Fort Bliss a major cavalry post, extending into the aftermath of the revolution.

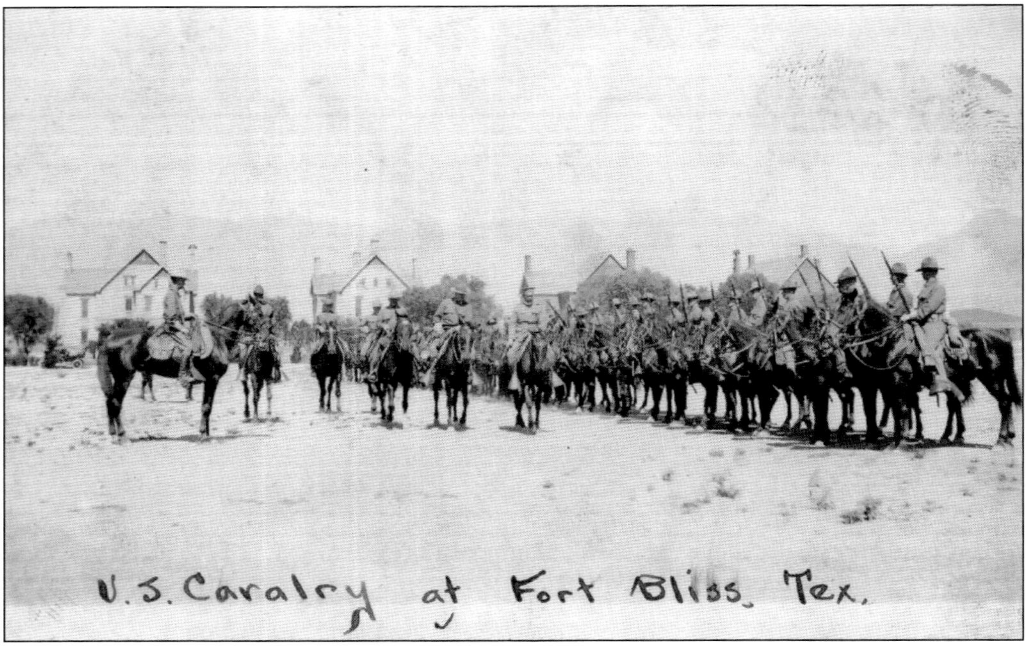

A cavalry review is staged for a visiting officer on the main Fort Bliss parade field. Four troops of the 4th Cavalry Regiment replaced companies of the 23rd Infantry in 1912, beginning a historic transition from a post that was primarily infantry before. Horse cavalry had the ability to patrol the rough terrain of the US–Mexico border during the Mexican Revolution.

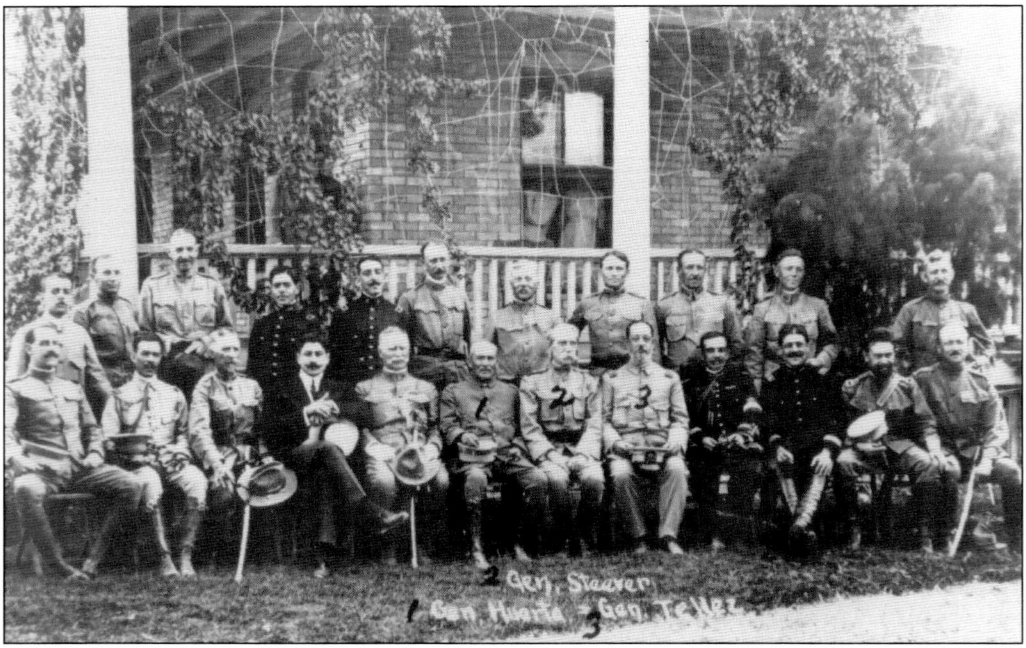

Brig. Gen. Edgar Zell Steever hosted Mexican commanders Victoriano Huerta and Joaquin Tellez at Fort Bliss in 1912. Steever commanded the El Paso Military District between 1911 and 1912. Huerta and Tellez had just defeated northern Mexican revolutionaries, including Pascual Orozco. Here, they are seated in front of Quarters 1, Steever's quarters. Huerta would rise and fall as Mexico's dictator and later be a prisoner at Fort Bliss.

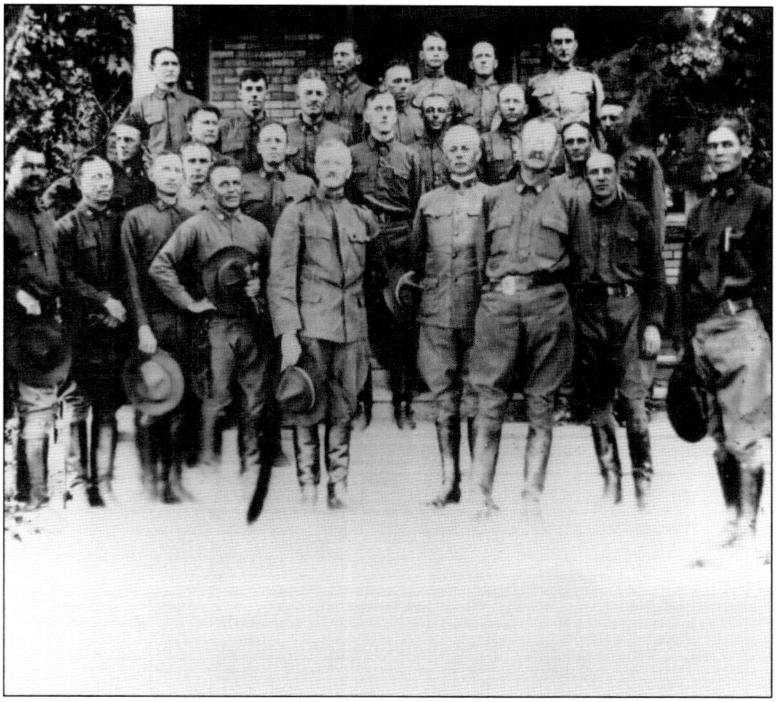

In 1914, the Army dispatched the 8th Infantry Brigade from California to Fort Bliss. Brig. Gen. John J. Pershing stands with his staff and commanders in front of Quarters 1 in 1914. Planning to have his family join him, on April 27, 1915, he learned that his wife and three daughters had perished in a house fire at the Presidio of San Francisco. Only his son Warren survived.

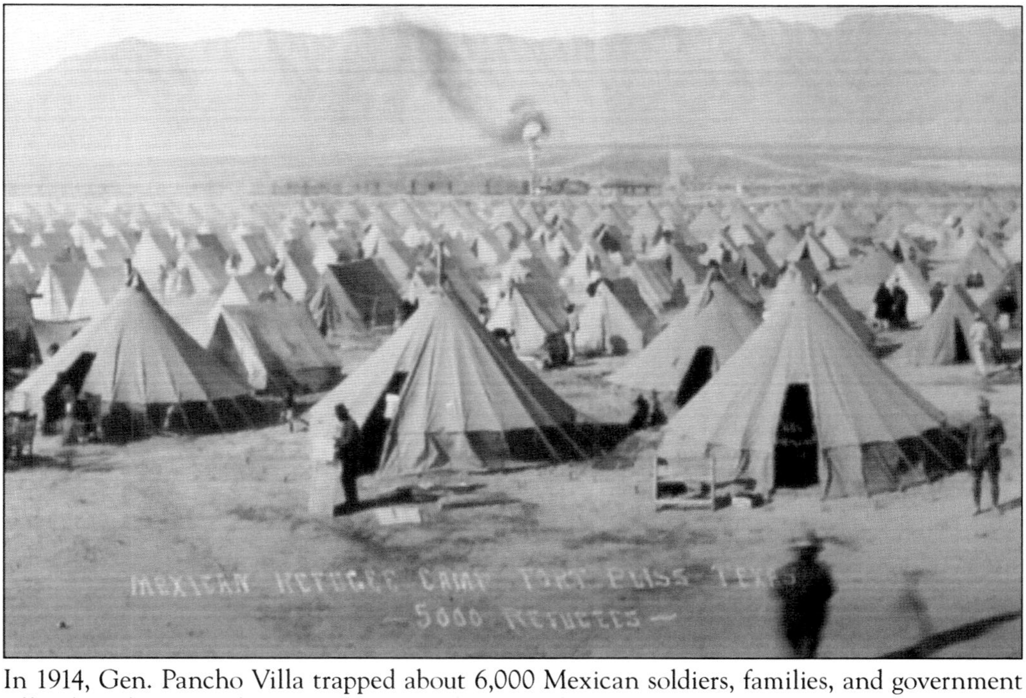

In 1914, Gen. Pancho Villa trapped about 6,000 Mexican soldiers, families, and government officials at the town of Ojinaga, Mexico. These people crossed the border and surrendered to the 15th Cavalry Regiment in Presidio, Texas. They were moved to Fort Bliss and interned for four months in a detention camp. The Army garrison at Fort Bliss guarded the camp until the people were moved to Fort Wingate, New Mexico, and released thereafter.

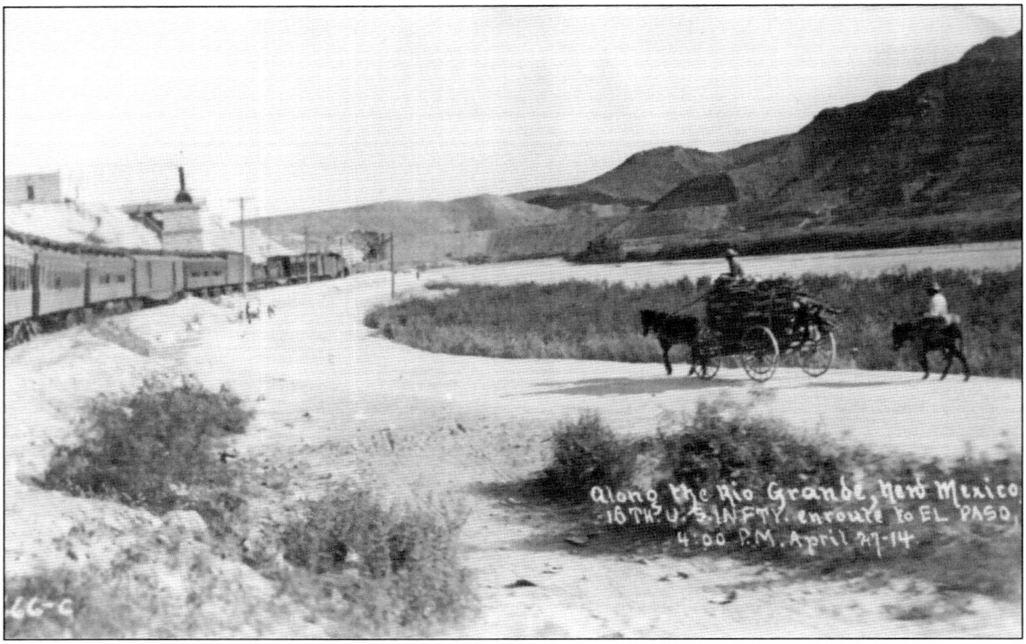

The 8th Infantry Brigade moved from the Presidio of San Francisco to Fort Bliss in April 1914. This included two regiments of infantry, the 6th and 16th. Rounding the bend of the Rio Grande at the smelter, 16th Infantry soldiers got a view of the border that they would soon secure from revolutionary incursions and smuggling. Their arrival overcrowded the post, and units occupied tent camps around El Paso.

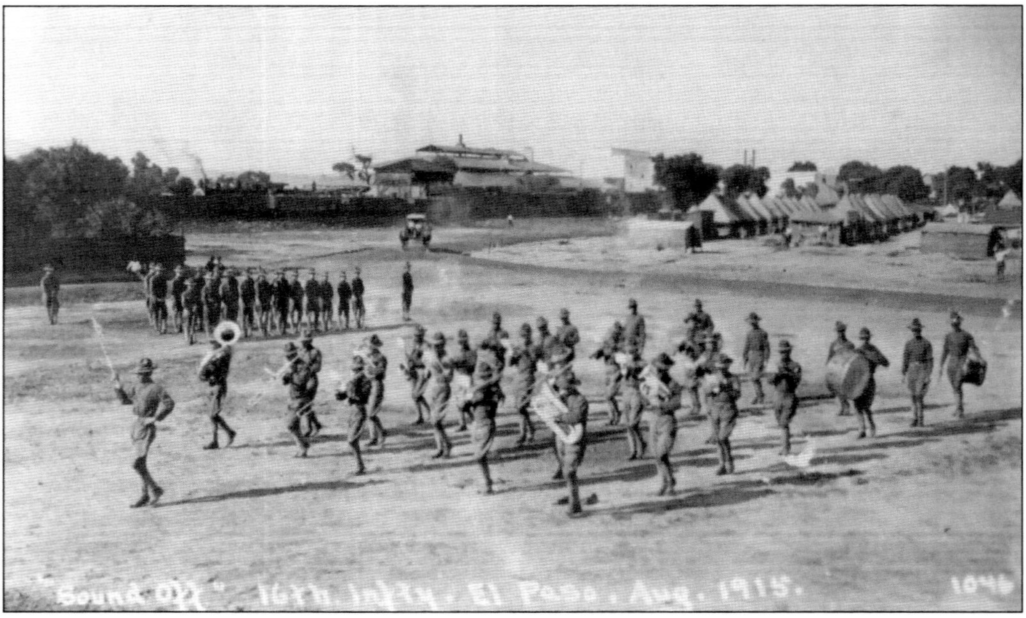

All Army regiments of the time had their own bands. Here, troops of the 16th Infantry stand at ease watching their band sound off at Camp Cotton in August 1915, near what is now the Chamizal Memorial. The International Brick Company in El Paso stands in the background. This was likely a retreat ceremony at day's end, with a guard company preparing to post guards around the camp.

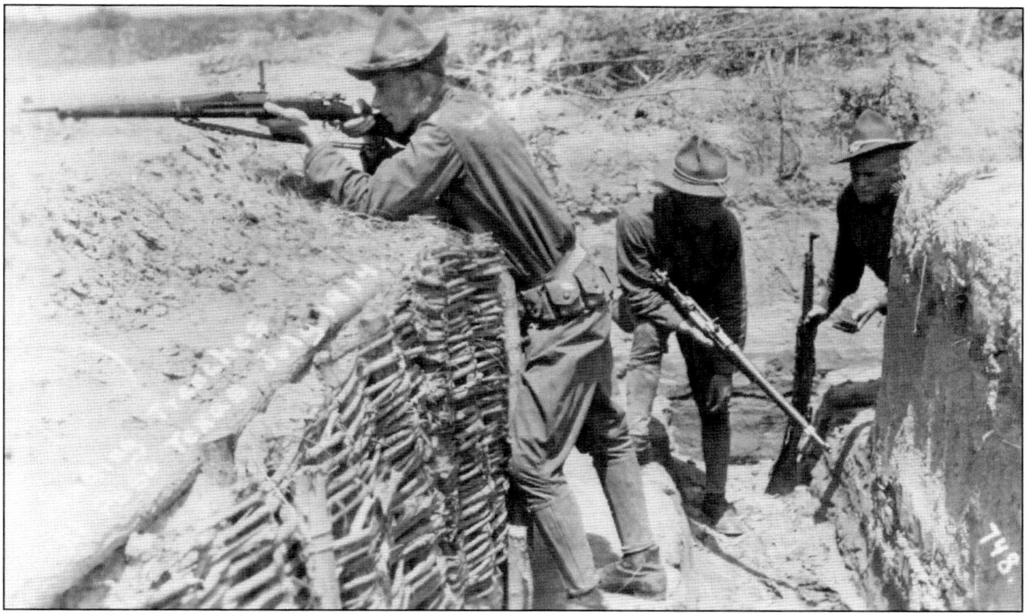

Troops guarded the border but also conducted training. Here, 16th Infantry soldiers train in a trench line, foreshadowing events in a few years in Europe. They train with the issue Model 1903 Springfield rifle, a bolt action repeater. This caliber .30-06 rifle had an effective range of up to 800 yards in the hands of an expert marksman.

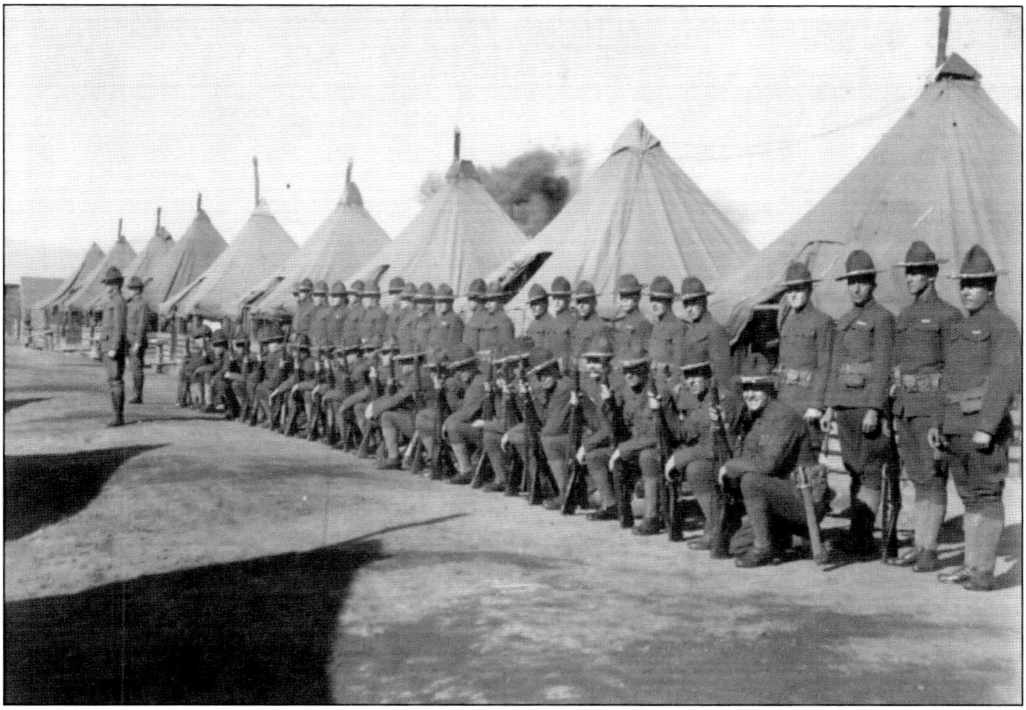

Company D, 16th Infantry pose for a photograph in 1914. They wear issue Model 1912 webbed cartridge belts and carry Model 1903 Springfield rifles with the M1905 16-inch bayonet. They also wear the olive drab wool uniforms with canvas leggings. They lived for months in these tents near the border—a rough life.

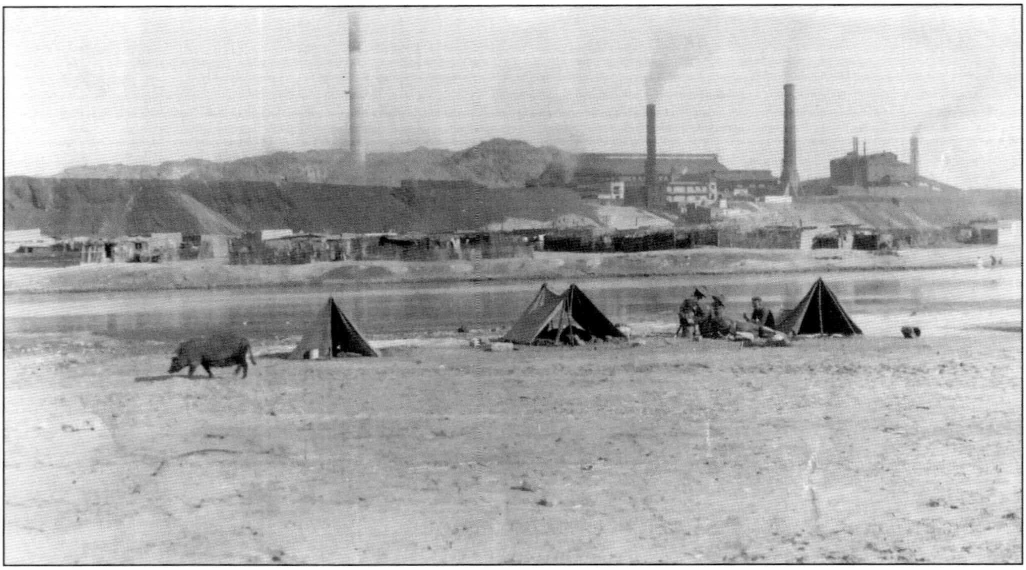

The arriving troops settled into a routine of guarding the border. These troops were camped on the river near Border Monument 1, opposite the smelter. A sow hog is wandering nearby, possibly from across the border. With the Mexican border a stone's throw away, troops could keep an eye out for any suspicious movements or threats.

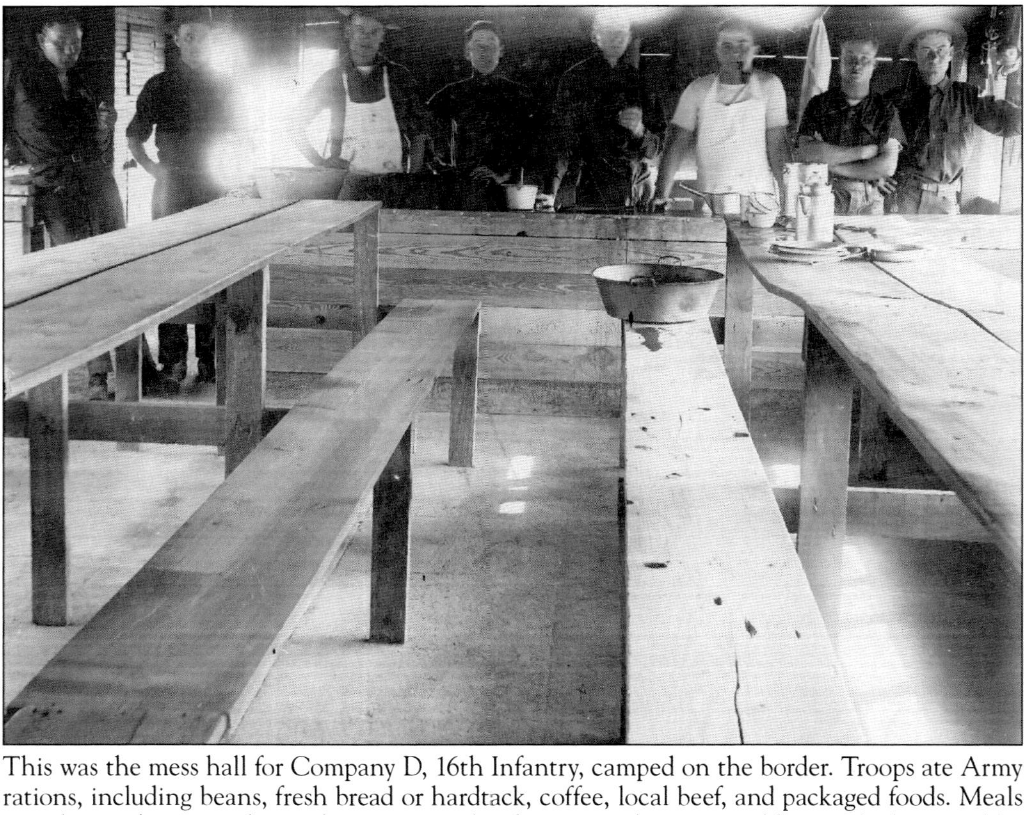

This was the mess hall for Company D, 16th Infantry, camped on the border. Troops ate Army rations, including beans, fresh bread or hardtack, coffee, local beef, and packaged foods. Meals were better due to modern refrigeration and icehouses, and troops could enjoy fresh vegetables and fruits when they were available.

American troops stand at Border Monument 1, across the Rio Grande from the smelter. The Mexican border is only a step or two away. Smuggling intensified along the border in El Paso and downriver, around Fabens and San Elizario. Despite the increased troop presence, smuggling was a brisk business.

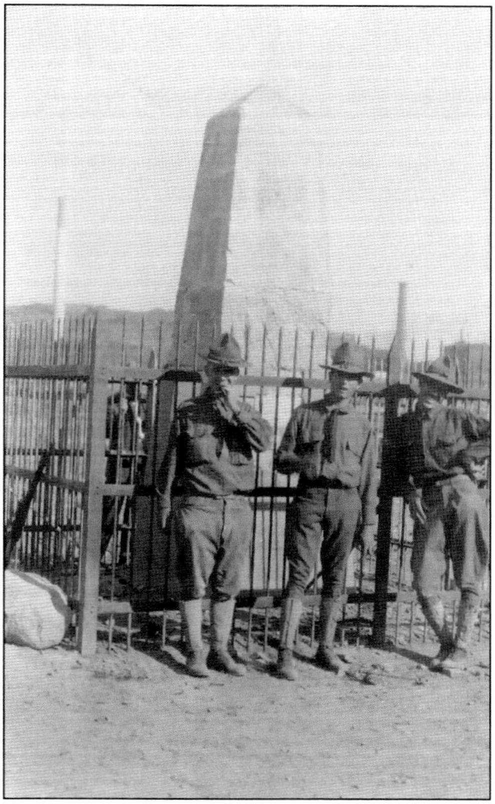

For troops not on duty, there were diversions in El Paso. The Army frowned on alcohol consumption, and it was always a problem, especially on payday. Here, troops clown in front of the wholesale liquor sign. Temperance was a major concern in the War Department, with Secretary Newton Baker decidedly against alcohol consumption by the troops.

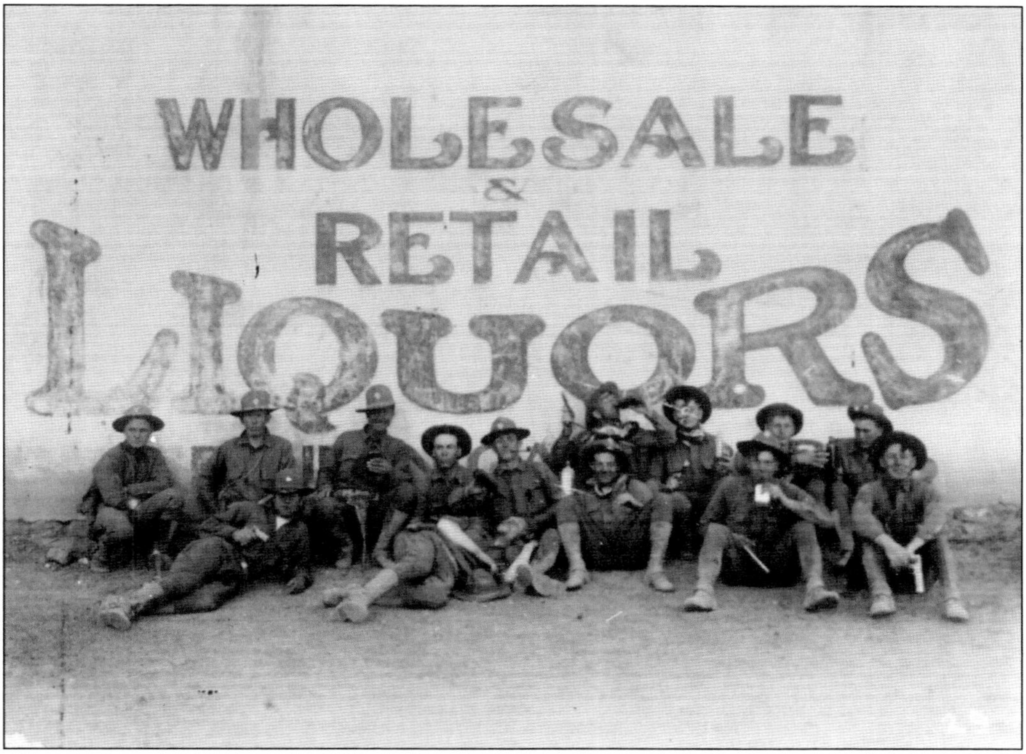

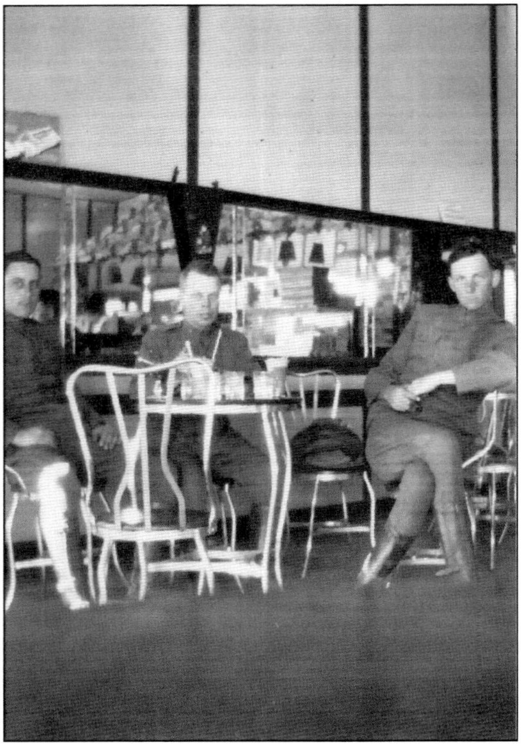

For other troops, there were more wholesome diversions. These troops enjoy ice cream at Elite Confectionery. This sweet shop was in the ground floor of the Buckler Building in downtown El Paso. Its previous customers included the revolutionaries Pancho Villa and Pascual Orozco when they were in exile in El Paso in 1911. Some of these troopers undoubtedly chased Villa in Mexico.

In August 1914, revolutionary leaders Álvaro Obregón and Pancho Villa met with Brig. Gen. Pershing in El Paso. The intent of this meeting was to discuss keeping the conflict away from the border, especially cities. Obregón and Villa were brought to Fort Bliss where they viewed cavalry demonstrations and had refreshments in Pershing's quarters. Villa's coat bulges where his pistols are likely concealed.

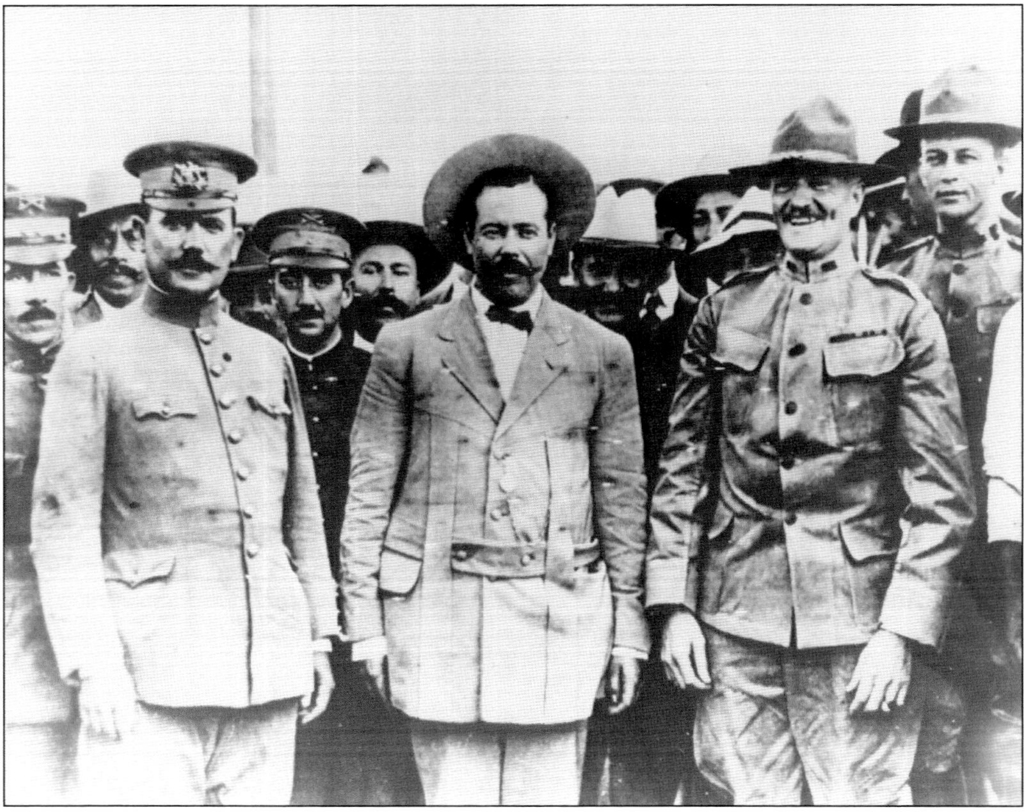

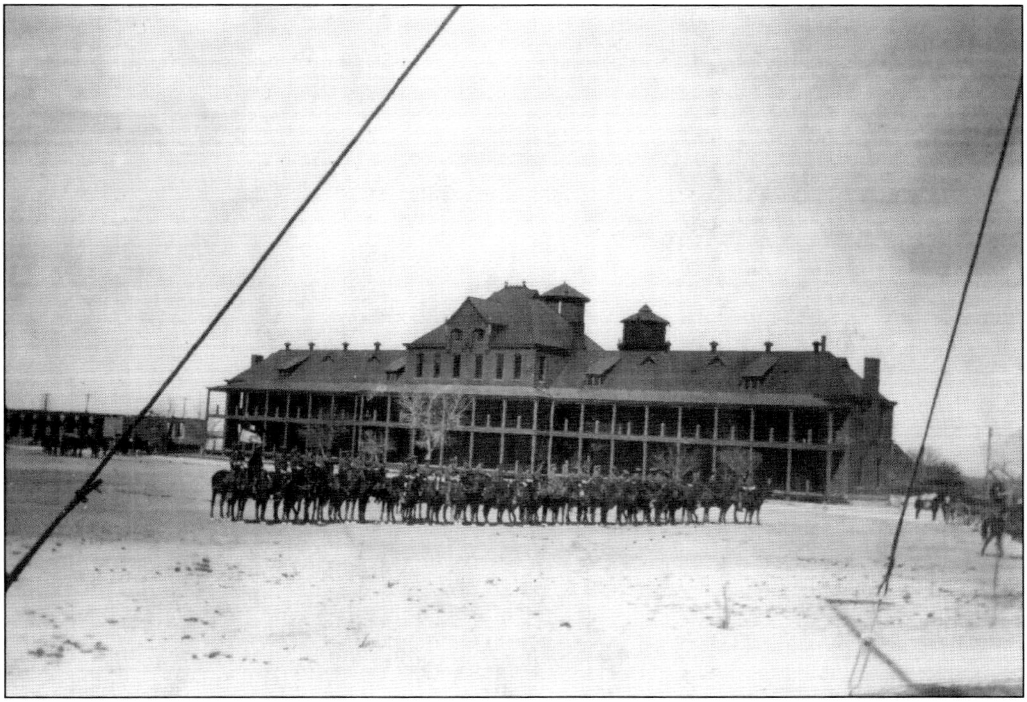

A squadron of the 15th Cavalry Regiment stands in line in front of what is now Building 113. To the left rear are new barracks under construction for cavalry. The cables to the right and left support the post flagpole. Infantry was suited to fixed guard posts along the border, but only horse cavalry could patrol the rough areas of the border to the east and west.

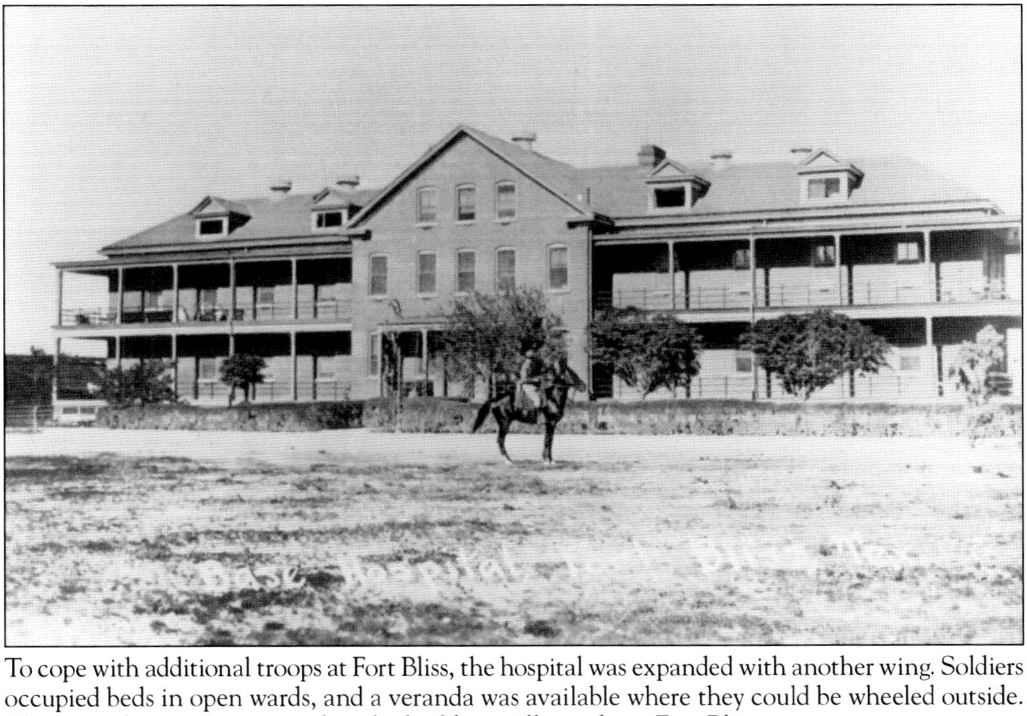

To cope with additional troops at Fort Bliss, the hospital was expanded with another wing. Soldiers occupied beds in open wards, and a veranda was available where they could be wheeled outside. These porches are now gone, but the building still stands on Fort Bliss.

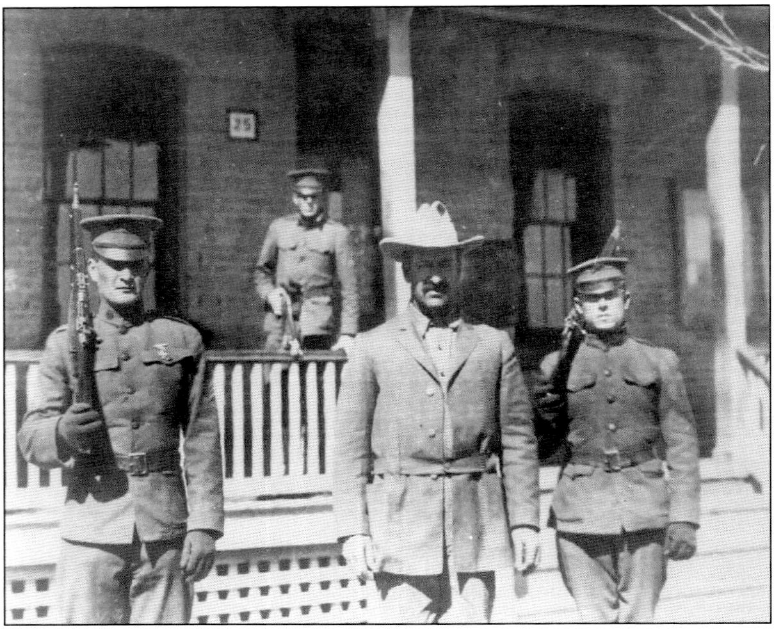

Mexican revolutionary Gen. Jose Ines Salazar stands between two guards in front of Building 25, the post guardhouse. Salazar was detained during the Mexican refugee affair and held pending charges of smuggling and violations of US neutrality laws. Salazar was acquitted in his trial. The guardhouse had jail cells in the basement, but these cells are now gone.

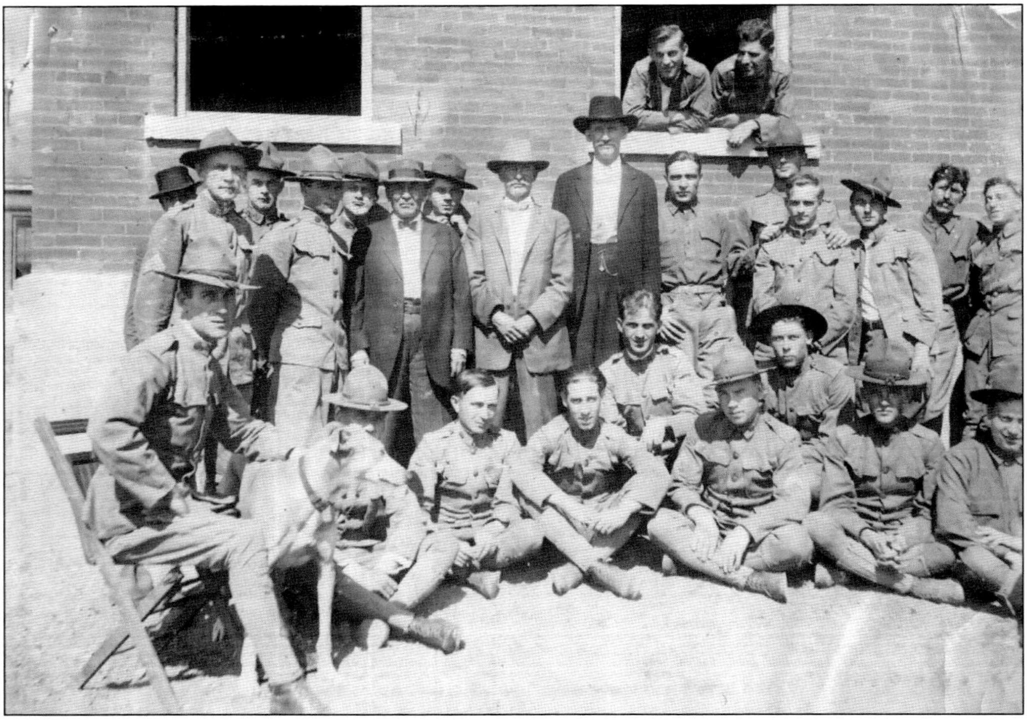

Former Mexican president Victoriano Huerta stands with guards and soldiers at Fort Bliss in 1915. Forced out of Mexico in 1914, Huerta conspired with supporters to retake power in Mexico. US law enforcement and cavalry soldiers detained him in Newman, New Mexico, in June 1915. He was placed in the Fort Bliss guardhouse until lawyers could arrange bail. A heavy drinker, Huerta died in El Paso in 1916.

Attracted by the possibility of action along the border, Lt. George S. Patton Jr. obtained an assignment to the 8th Cavalry Regiment at Fort Bliss in 1915. He stands in front of 222 Sheridan Road, where he lived with his wife and two daughters. After the Columbus Raid of 1916, Patton managed to join the Punitive Expedition, where he gained fame in a shoot-out with Villistas, Villa's band of militants.

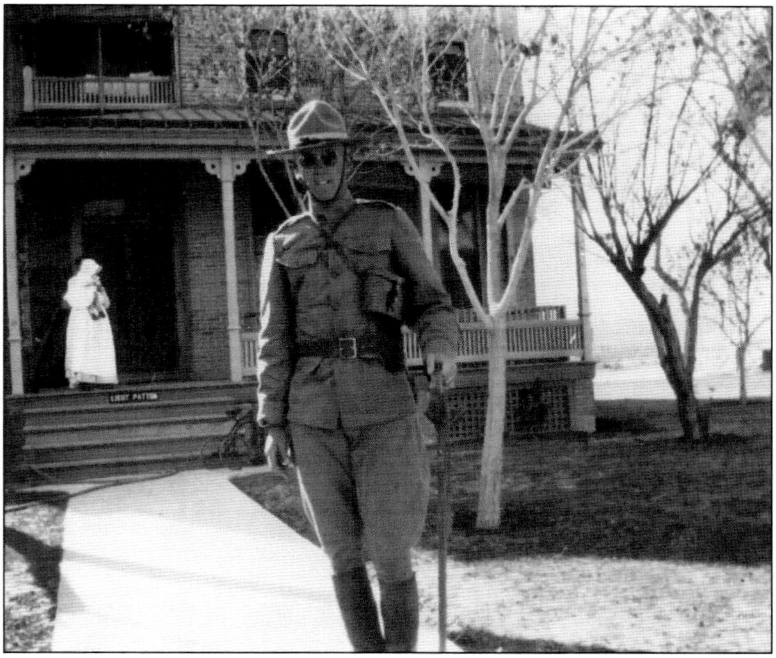

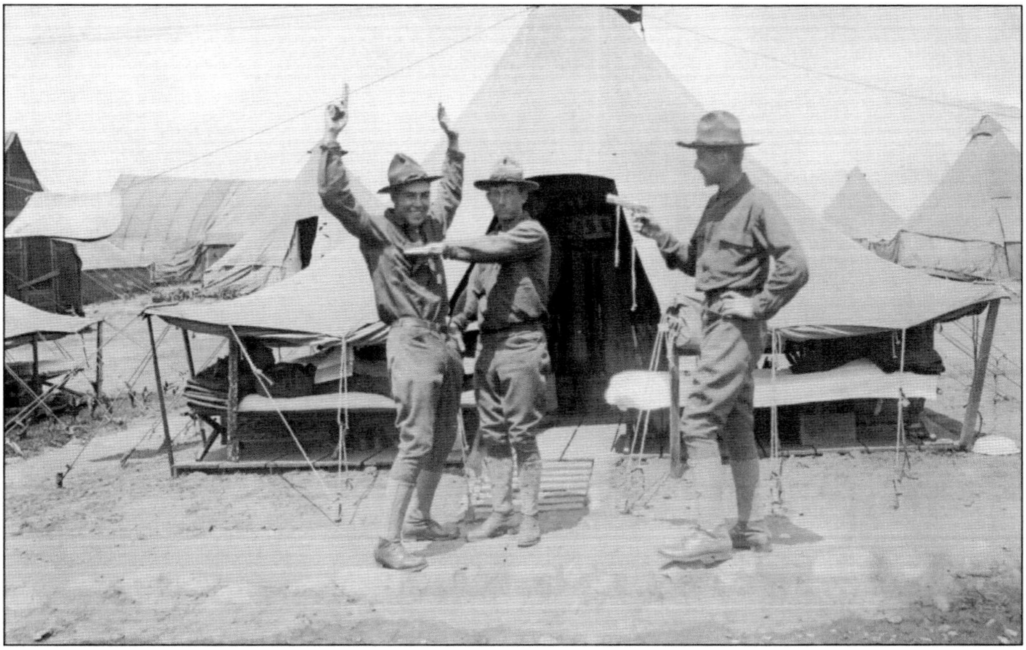

The raid on Columbus, New Mexico, in 1916 finally forced action by the United States. To confront this incursion, Pres. Woodrow Wilson ordered a punitive expedition into Mexico to break up Pancho Villa's forces and capture Villa. Fort Bliss was mobilized as a major support base, which brought in more cavalry. This picture shows cavalrymen clowning in a border camp in 1916.

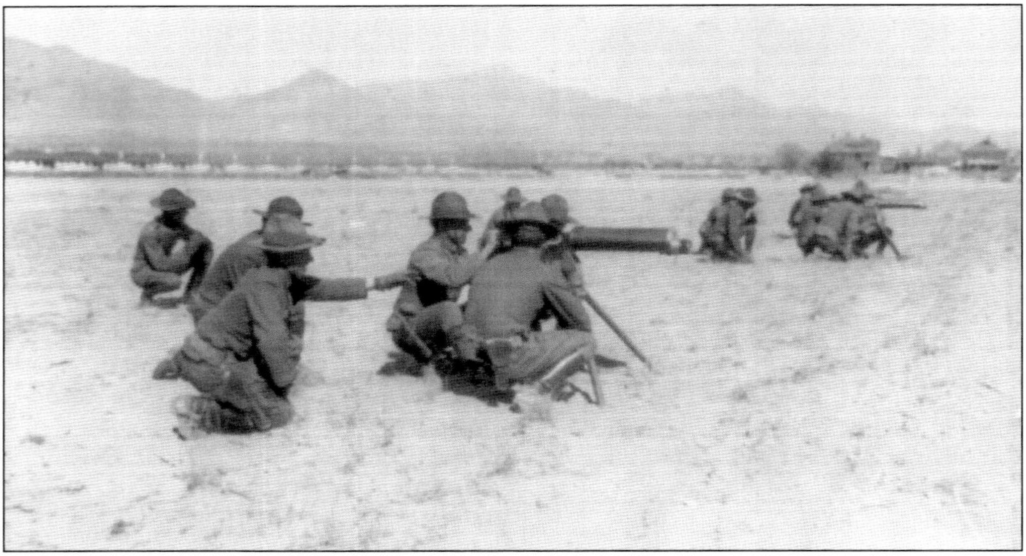

Many innovations appeared in the Army of the 20th century. The Army was testing and employing machine guns. Here, soldiers of the 8th Cavalry fire a M1915 Colt-Vickers machine gun east of the post. This gun was heavy but very reliable. The Army ordered 4,000 of these guns and used them during the Punitive Expedition and after.

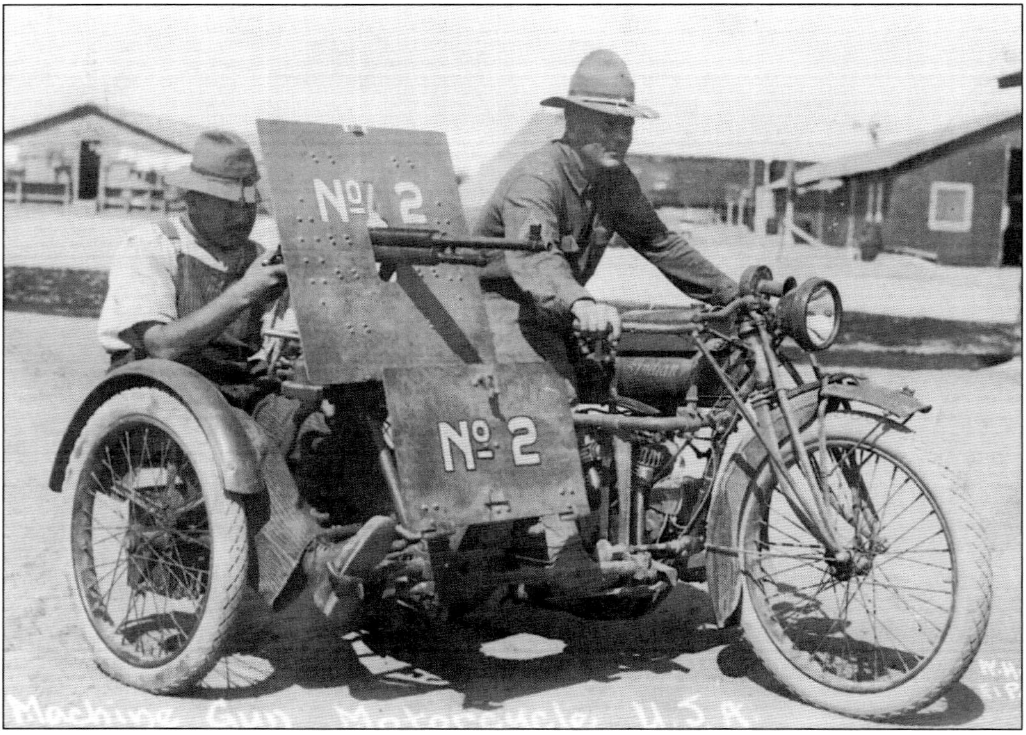

A cavalry trooper sits astride an Indian motorcycle with a sidecar-mounted Benet Mercier machine gun. The Army was just beginning to evaluate motorized vehicles for use in military operations. The gunner appears to be a civilian. The consensus among cavalrymen at the time was that mechanized vehicles could never replace the horse in rough terrain. But subsequent events would prove that invalid.

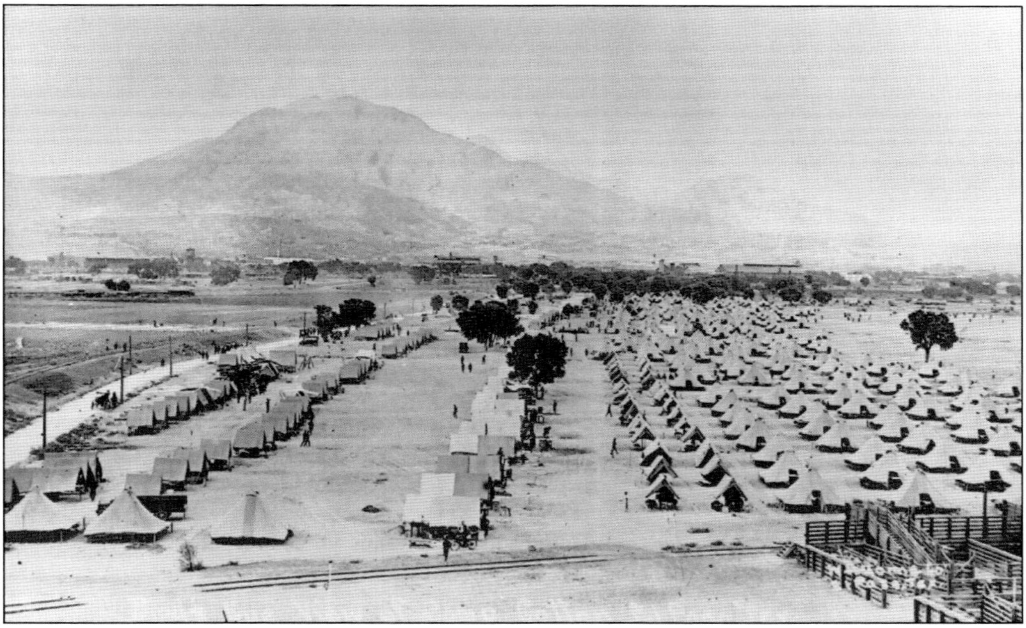

With the Punitive Expedition in Mexico and the remainder of troops patrolling the border, Fort Bliss was relatively empty for a short time. This would change when Mexican revolutionaries raided Glenn Springs, Texas, in May 1916. The president called up the National Guard for border duty, which added to troops at Fort Bliss. Tent camps sprang up around El Paso. This tent camp at Camp Cotton was one of several.

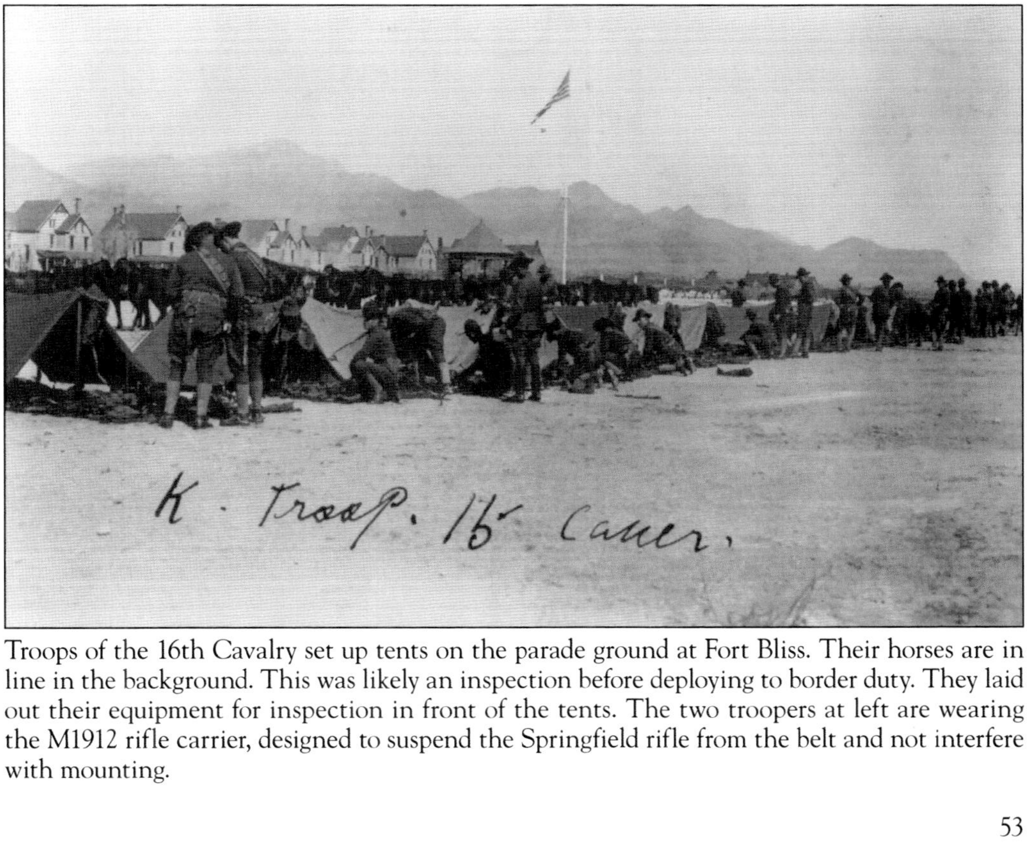

Troops of the 16th Cavalry set up tents on the parade ground at Fort Bliss. Their horses are in line in the background. This was likely an inspection before deploying to border duty. They laid out their equipment for inspection in front of the tents. The two troopers at left are wearing the M1912 rifle carrier, designed to suspend the Springfield rifle from the belt and not interfere with mounting.

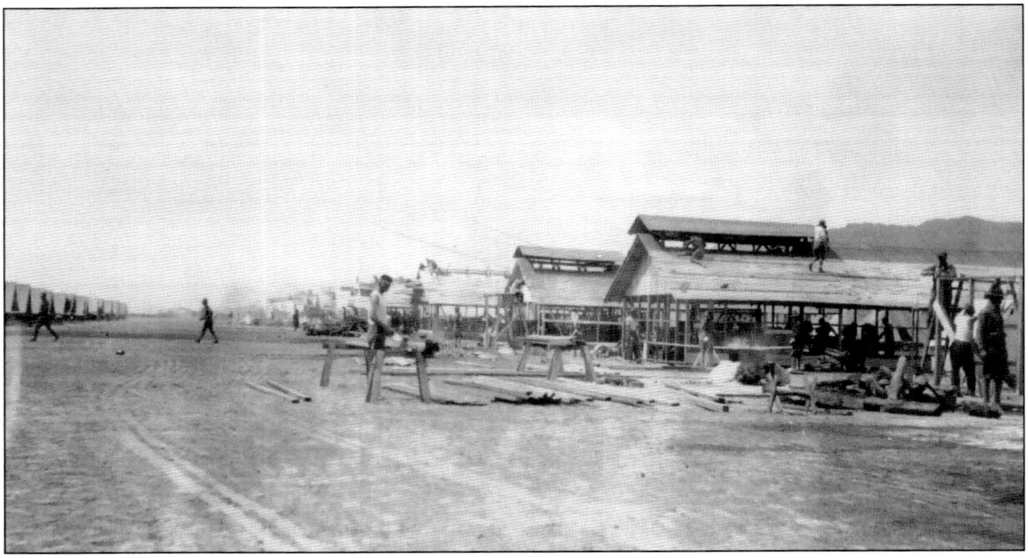

The Army quickly went to work producing wooden barracks and other buildings for deployed National Guardsmen. Some of this was done via contract, but a lot of the construction used troop labor. The sooner this could be done, the better. Until then, troops lived under canvas.

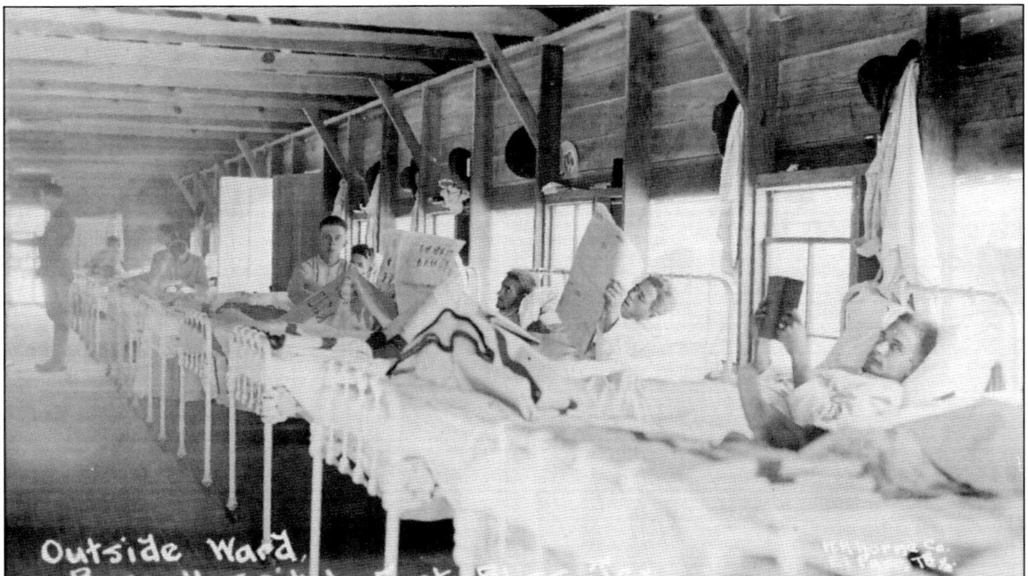

Troopers who were sick or wounded had this as a hospital. The main post hospital was too small to handle the influx of troops from the general mobilization, so some of the temporary wooden buildings had to serve. Medical care was reasonably good—much better than during the Civil War. But the great influenza epidemic of 1918 was yet to come.

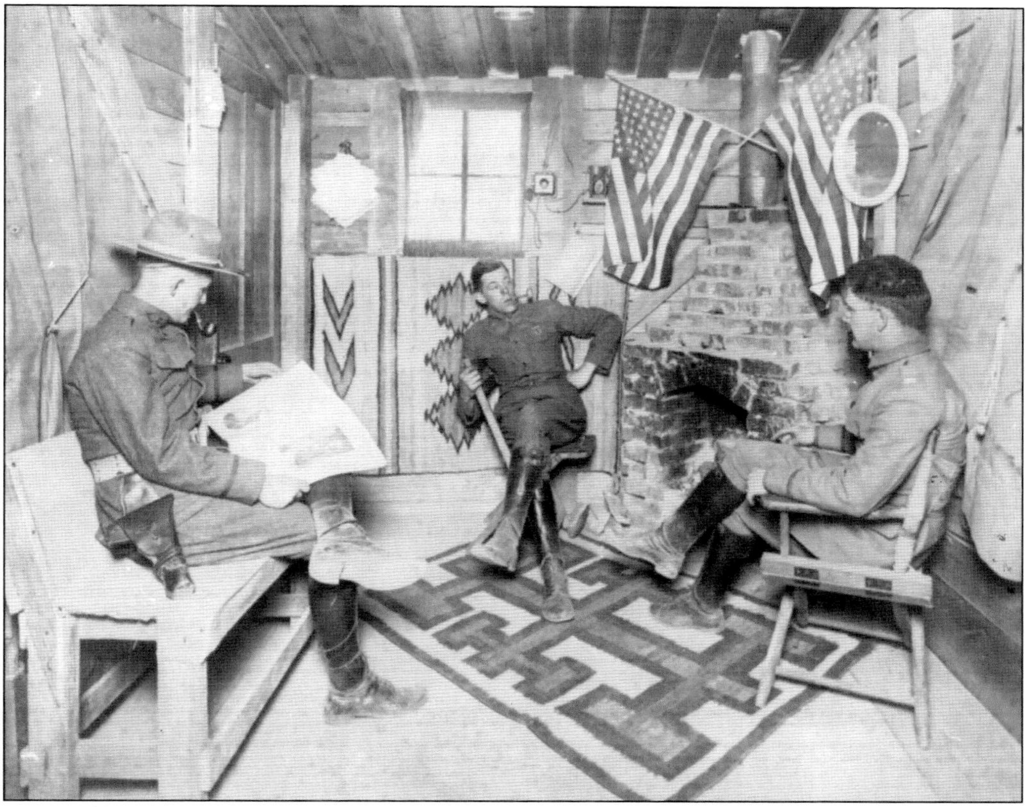

Junior officers relax and read in their quarters. They share a brick fireplace, rough-hewn benches and chairs, and rugs that were probably bought locally, likely in Lynchville. The mobilization for the Punitive Expedition proved to be a dress rehearsal for the United States' entry into World War I in less than a year's time.

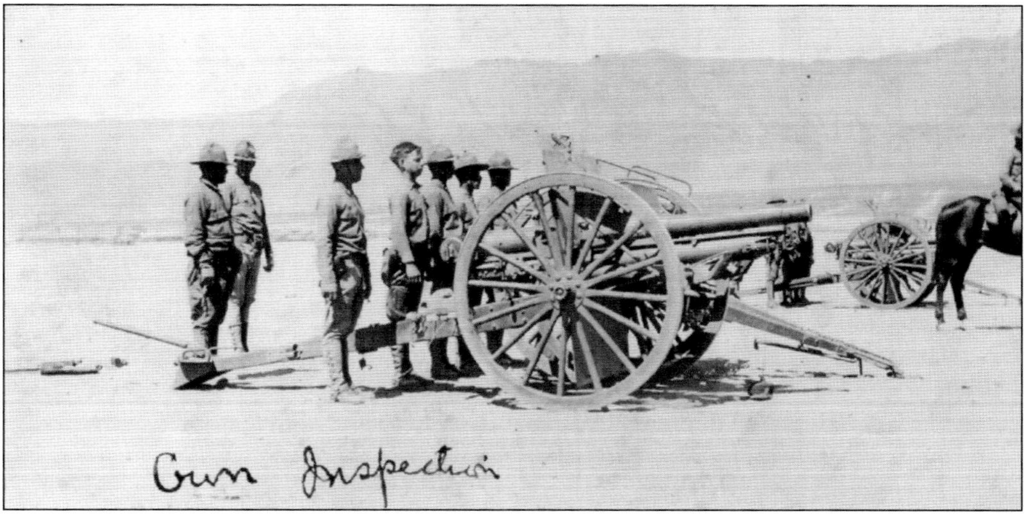

The Army brought field artillery into Fort Bliss to support the infantry and cavalry. The M1902 howitzer was horse-drawn and could give a good account of itself in battle. A battalion of these guns accompanied the Punitive Expedition into Mexico, but they were not used. The gun had a range of 6,500 yards, which would prove important in El Paso in the future.

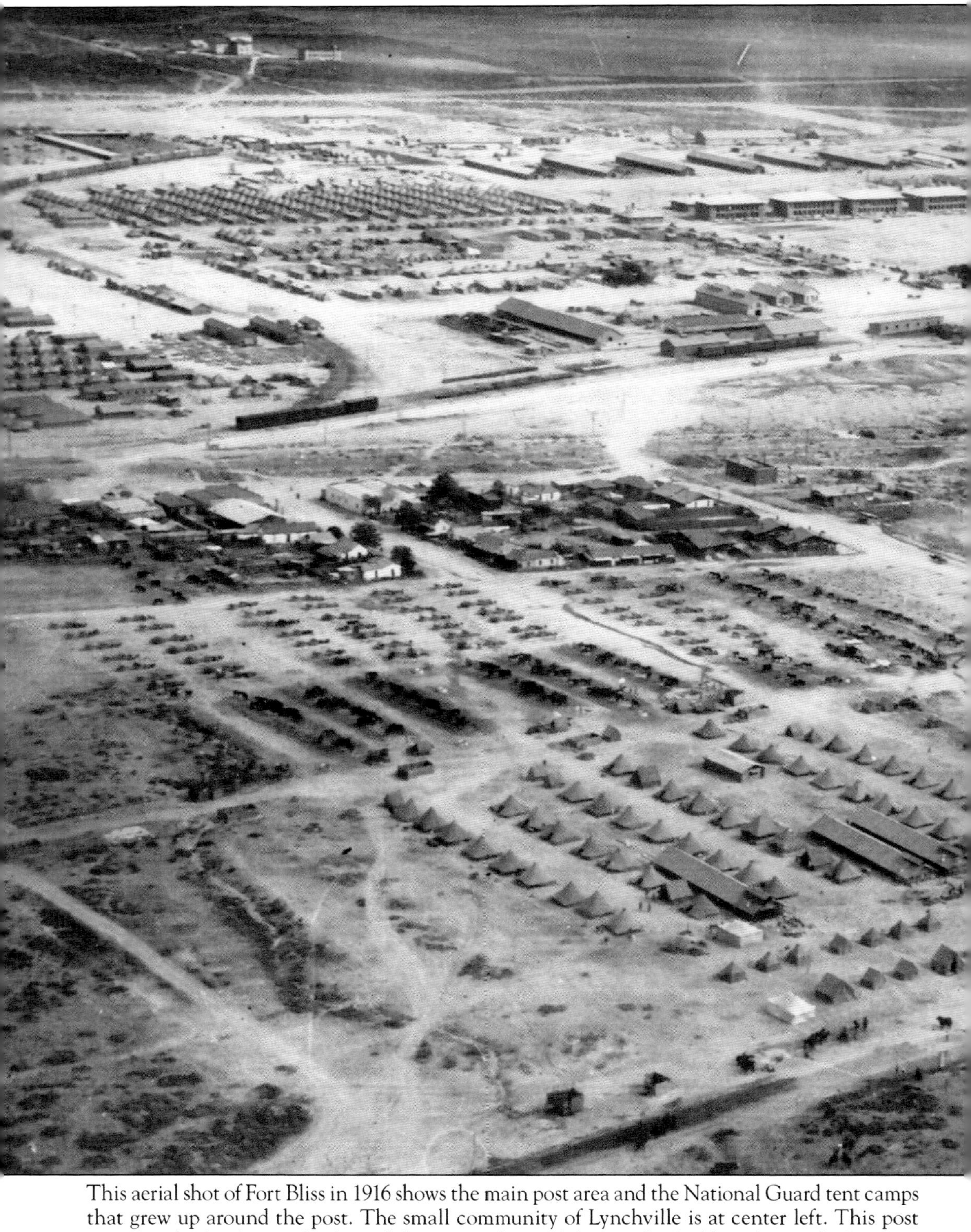

This aerial shot of Fort Bliss in 1916 shows the main post area and the National Guard tent camps that grew up around the post. The small community of Lynchville is at center left. This post served as the focal point for the mobilization and served thousands of soldiers during the Punitive Expedition. The post looked far different than the isolated 1893 installation. To the east of the

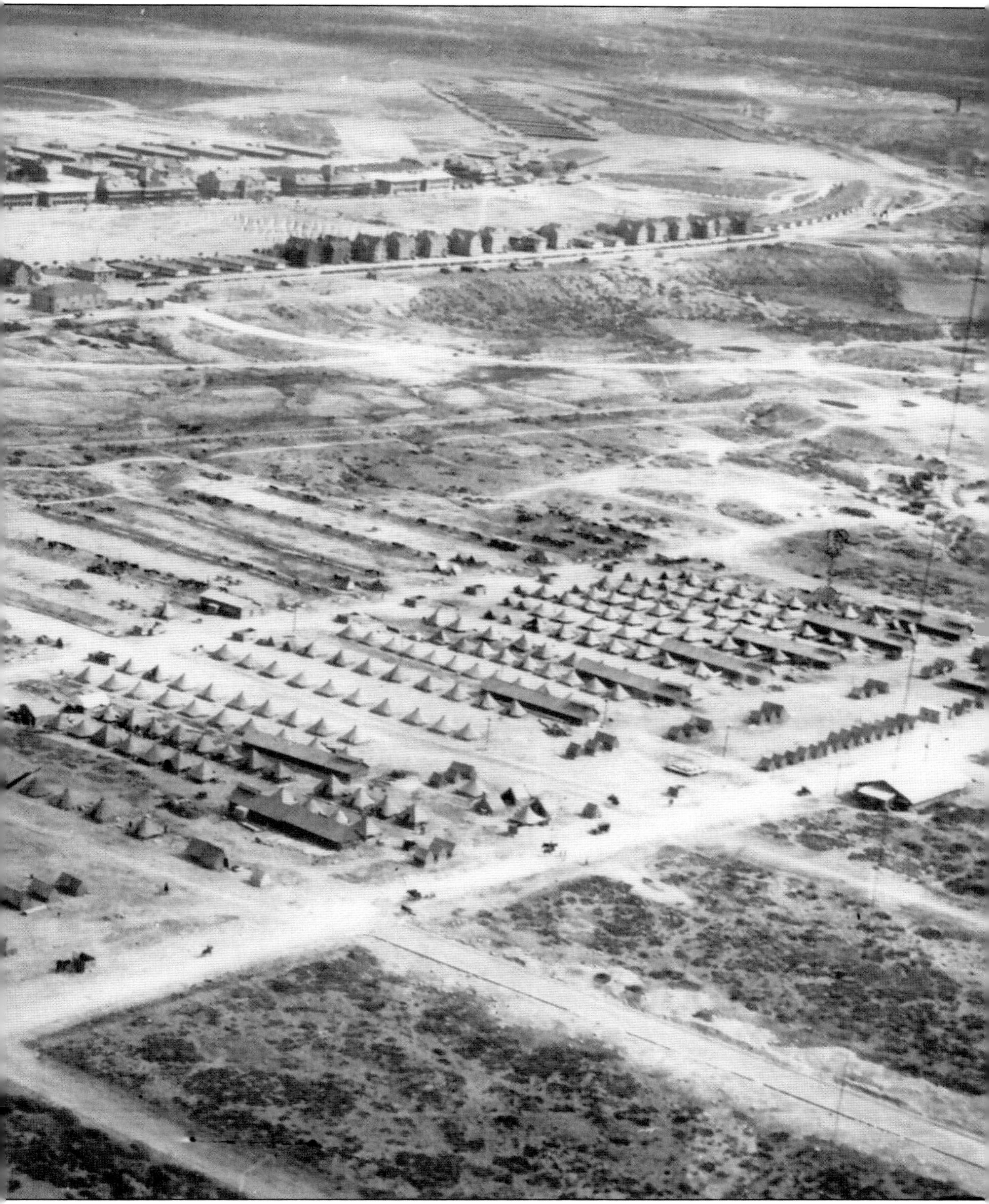

1893 barracks, the post has erected horse stables and support sheds. Tent camps for additional troops have been set up to the west and north of the post. In the upper left corner of the picture are the buildings of the El Paso School of Mines, soon to be transformed into an airfield.

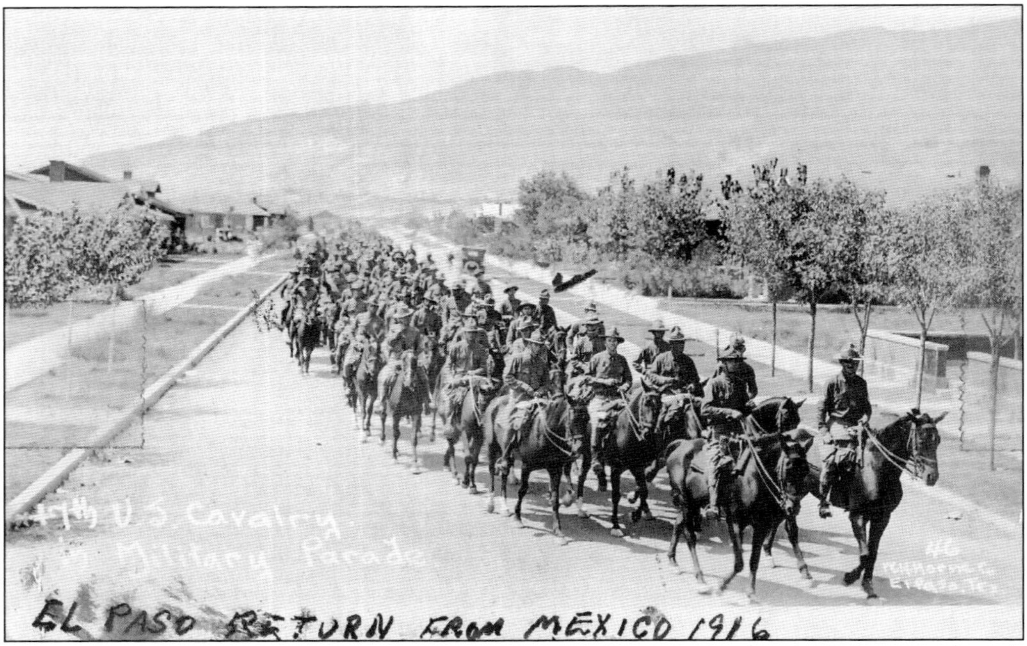

Troops returning from the Punitive Expedition moved onto Fort Bliss for billeting. The 7th Cavalry Regiment was one of those units to come to Fort Bliss. Here, they march to Fort Bliss down a street west of the post. As with the National Guard troops, this regiment would occupy rough quarters.

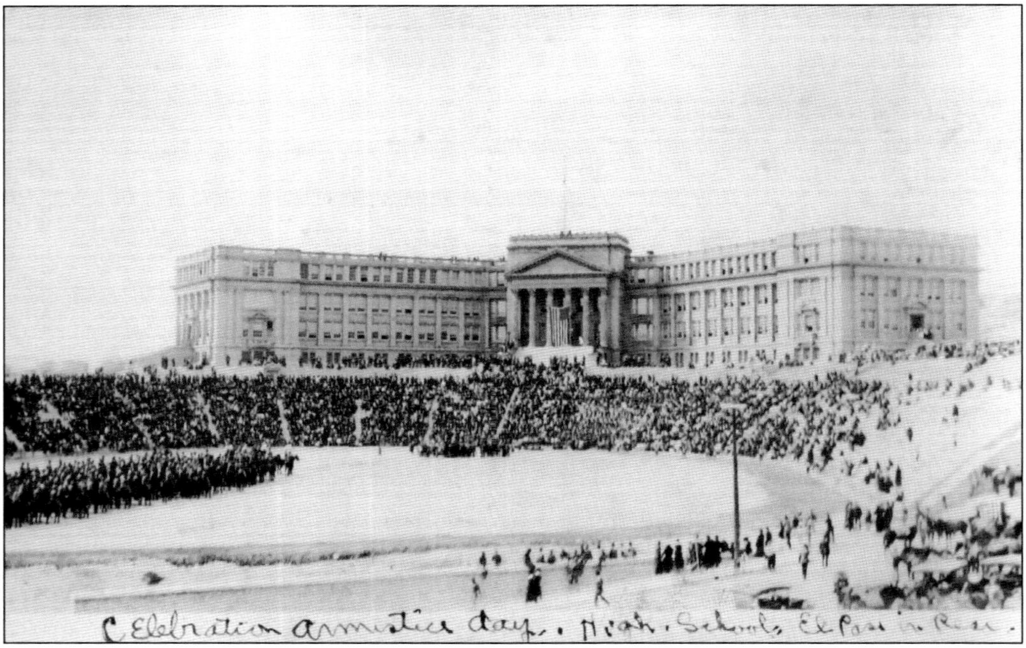

Fort Bliss served as a recruiting and training base during World War I. Cavalry troops stayed on to patrol the border, with only one cavalry regiment deploying to France. In 1918, El Paso and Fort Bliss celebrated Armistice Day at El Paso High School, with the cavalry much in evidence. But conflict on the border did not end.

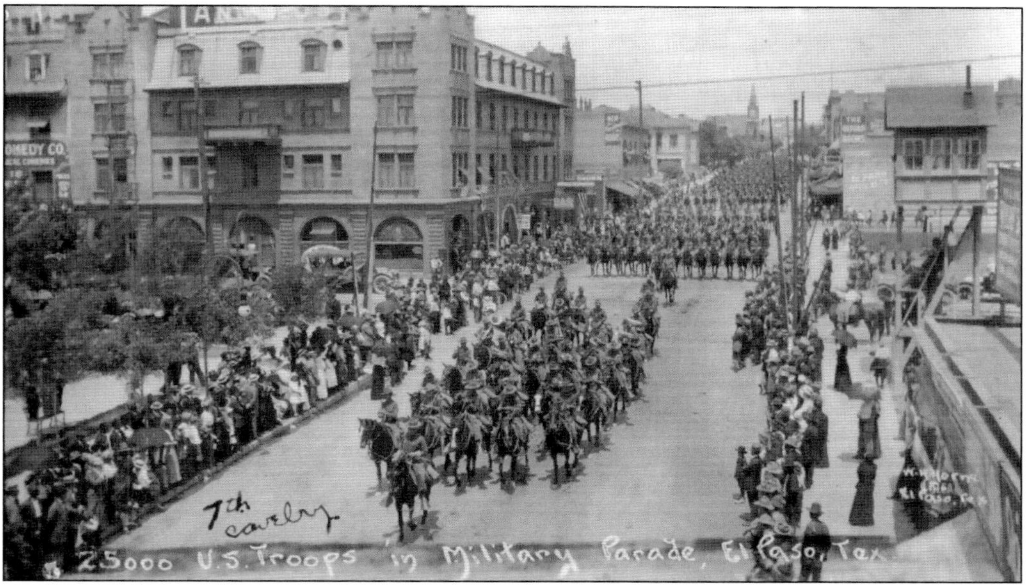

The citizens of El Paso viewed many parades of troops who had served in Mexico. Troops presented a military presence in El Paso for many years after the war. El Paso had become a major garrison town because of the Punitive Expedition and the border security mission that endured. And one more combat action would occur as the Mexican Revolution came to an end.

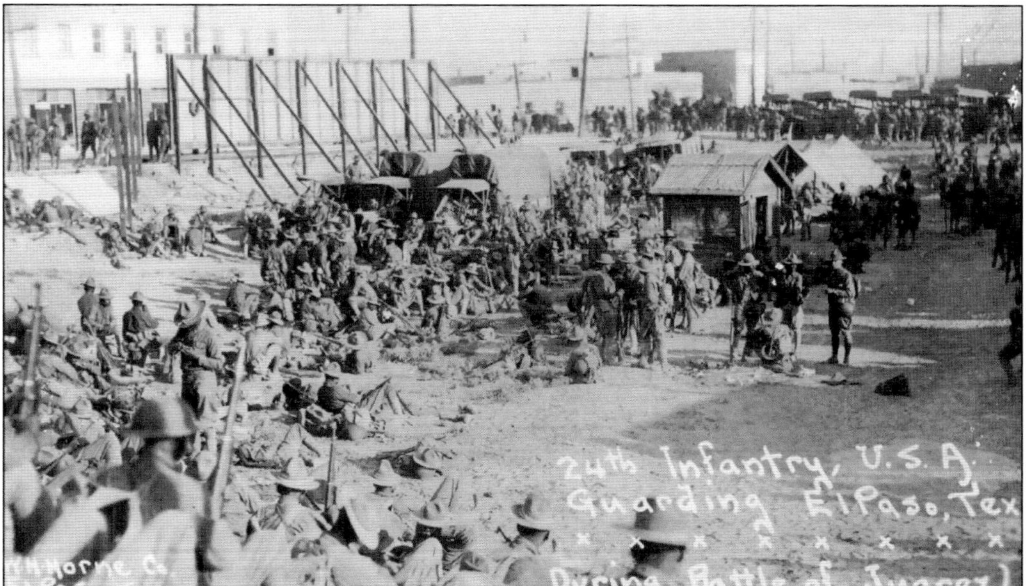

After World War I ended, Pancho Villa managed one last attempt to seize Cíudad Juárez in June 1919. African American troops of the 24th Infantry Regiment assembled near the international bridge in downtown El Paso, preparing to enter Mexico to counter Villa's attack. Troops are carrying the M1917 steel helmet, and motorized trucks are present. This was to be a combined arms operation, with infantry, cavalry, artillery, and combat engineers.

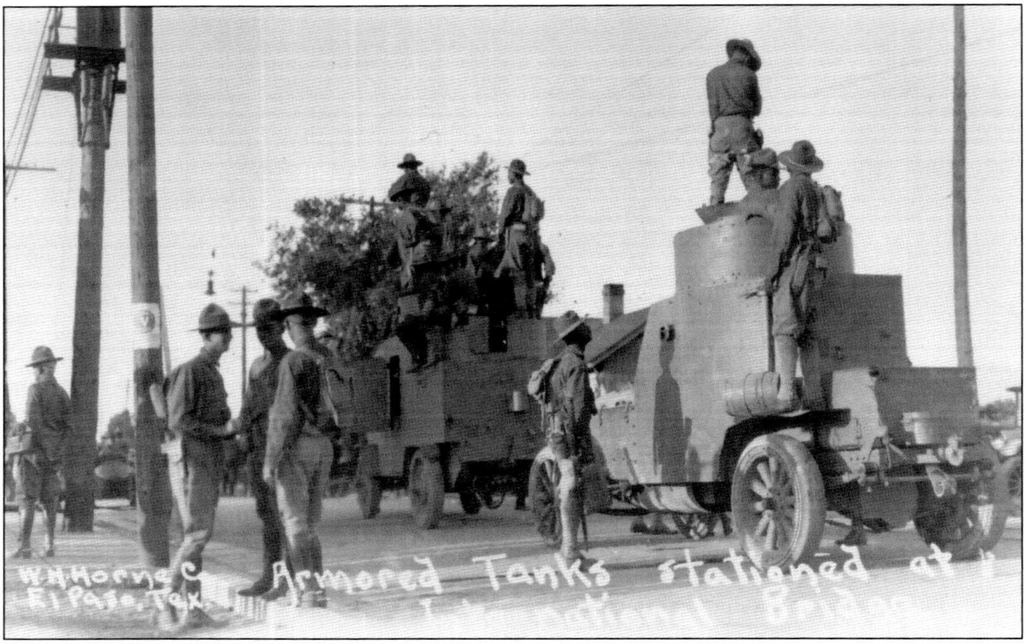

The commander at Fort Bliss, Brig. Gen. James Erwin, was well prepared to counter Villa's attempt. Here, African American troops of the 24th Infantry Regiment prepare to cross the international bridge into Mexico. Experimental Jeffery and White armored cars roll toward the bridge. Along with this incursion, two cavalry regiments crossed the Rio Grande downstream. Field artillery moved to fire on Villista concentrations in Juárez.

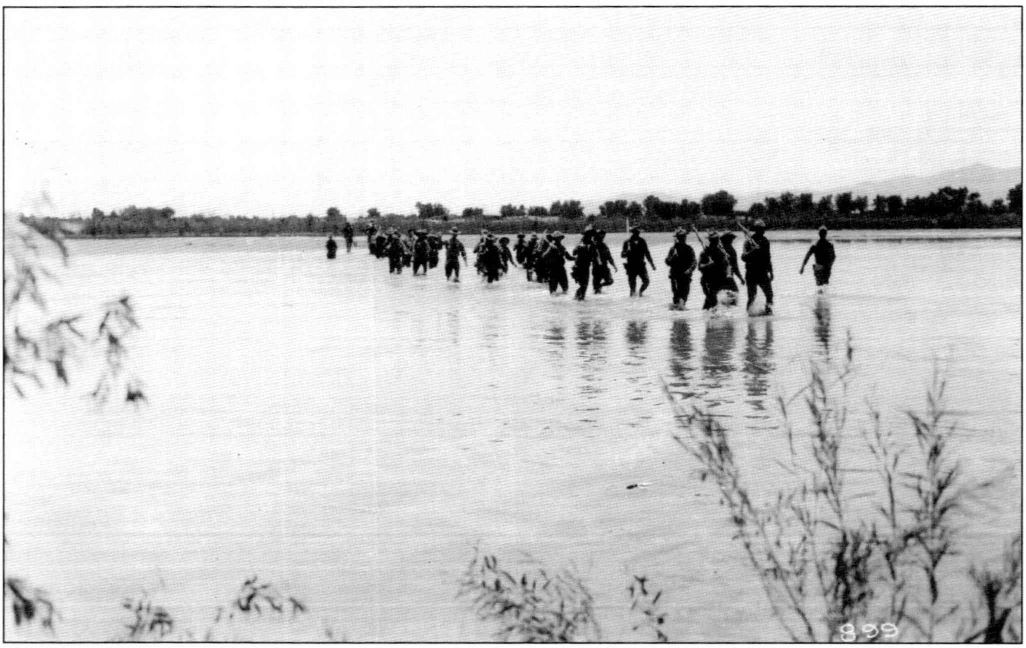

Infantry wade the Rio Grande in the June 1919 border incursion. These troops were determined that Villa would not seize Juárez. The plan was to fix the Villistas in place while the cavalry encircled them and captured them. American troops missed capturing Villa in the Punitive Expedition. This time, they were determined to succeed.

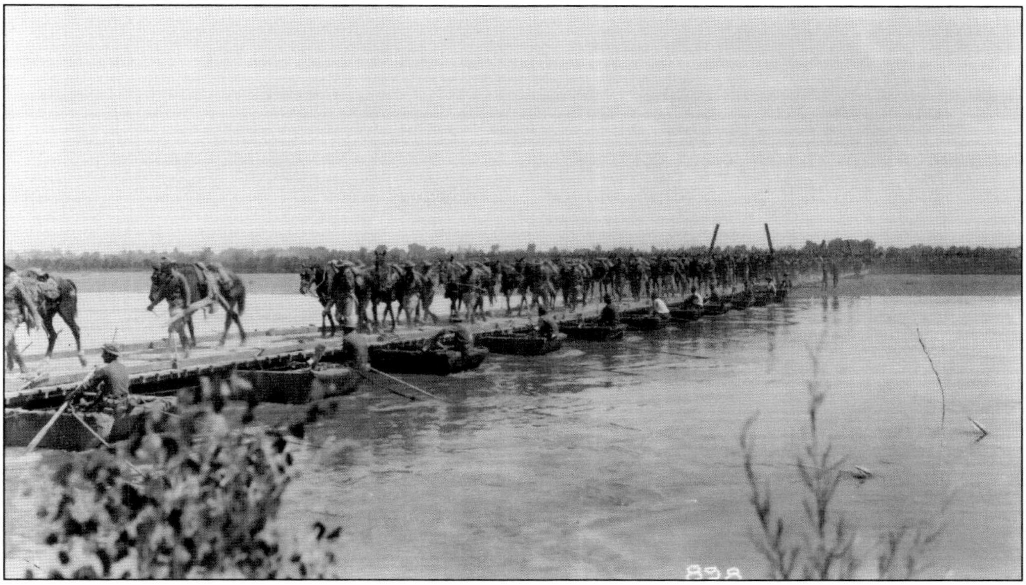
Cavalry troopers walk their horses back across the Rio Grande on June 16, 1919. It had been a near-run effort. Villa, seeing the danger of encirclement, managed to withdraw south into Mexico. Once again, the troops missed capturing the raider of Columbus, New Mexico, three years before.

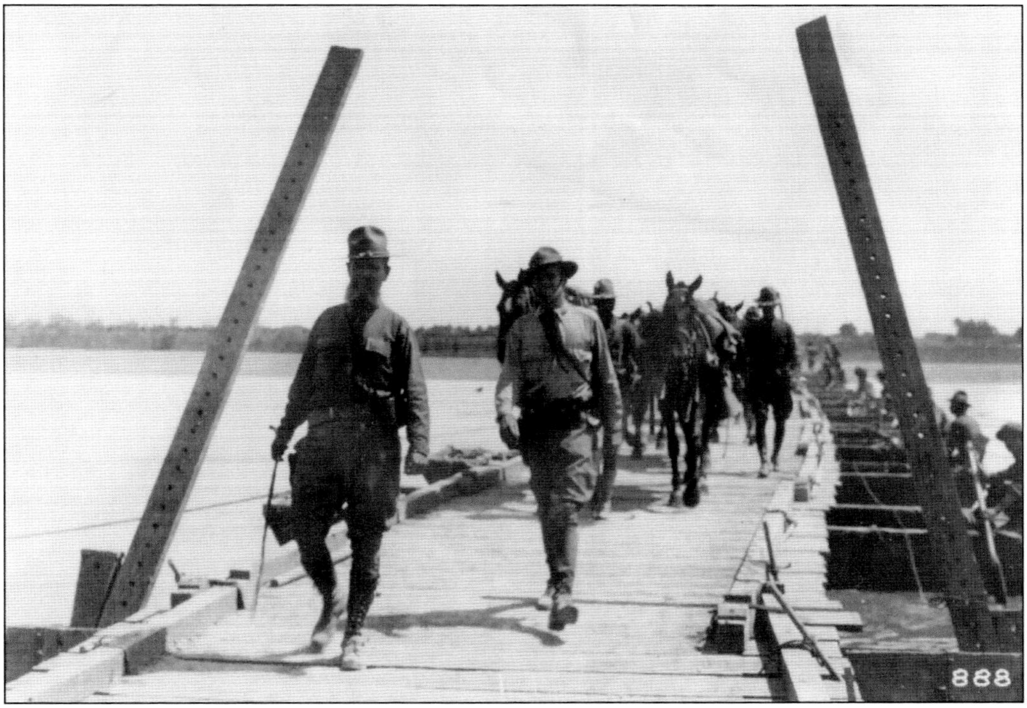

Cavalry brigade commander Col. Tommy Tompkins crosses a wooden pontoon bridge from Mexico. His troops, the 5th and 7th Cavalry, had missed collecting Villa and his staff. To his left is Captain Cunningham, his adjutant. Although they pursued Villa for 15 miles, once again he eluded capture. It was a bitter pill, but El Paso had been spared damage and Juárez remained in Mexican government hands.

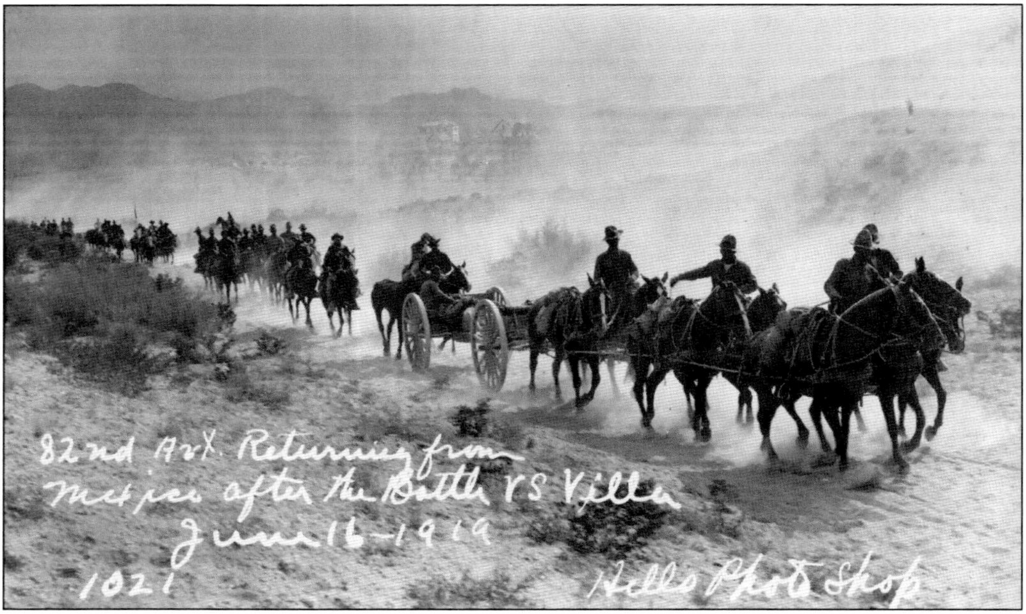

Troopers of the 82nd Field Artillery whip their teams to return to Fort Bliss. The guns had done their job in providing fire support for the American infantry and cavalry that had crossed the border. The Mexican federal commander in Cíudad Juárez was swift in filing a protest with the US government for the incursion of troops. But that incursion probably saved him from a Villista firing squad.

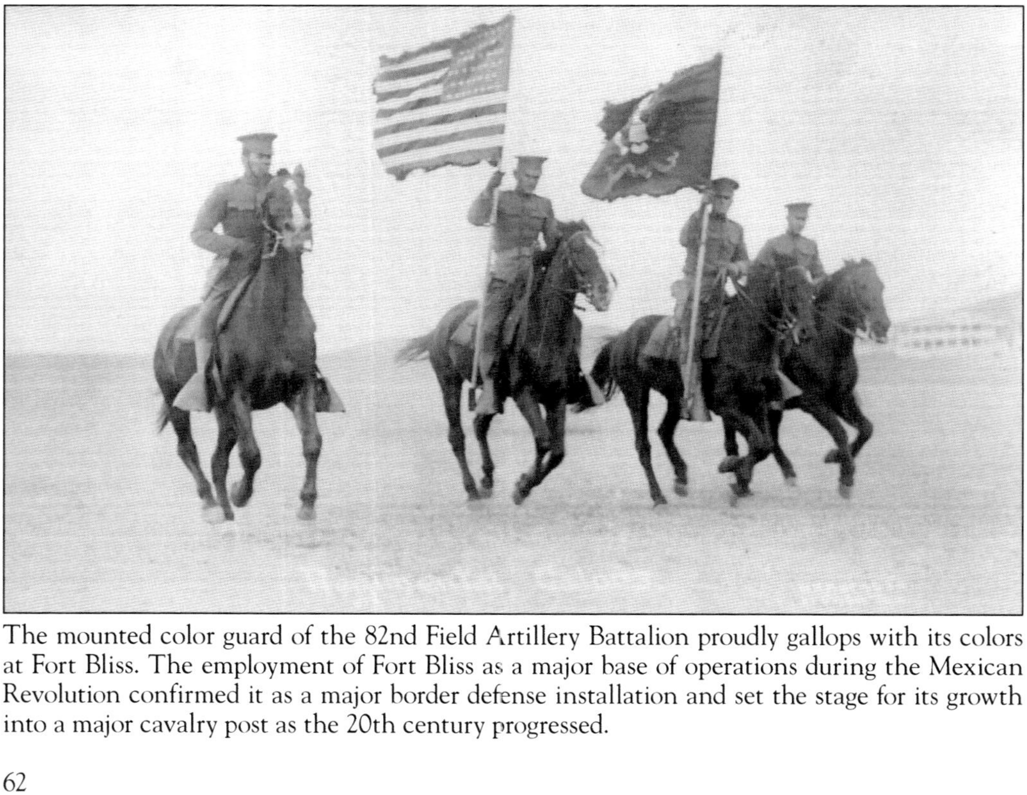

The mounted color guard of the 82nd Field Artillery Battalion proudly gallops with its colors at Fort Bliss. The employment of Fort Bliss as a major base of operations during the Mexican Revolution confirmed it as a major border defense installation and set the stage for its growth into a major cavalry post as the 20th century progressed.

Four

Horse Cavalry

The Cavalry Post

The Mexican Revolution was the catalyst that brought real growth to Fort Bliss. After acting as one of many frontier posts from the 1840s to the 1890s, the small post had to expand to accommodate the influx of troops that the revolution had precipitated. The pressure on the southern border brought thousands of Army National Guard troops to El Paso, most of which camped around the post. This proved to be somewhat of a dress rehearsal for World War I mobilization, but Fort Bliss did not serve as a major training post for that war.

After World War I ended, Fort Bliss resumed its service as a cavalry post. The cavalry regiments stationed at the post were consolidated into the storied 1st Cavalry Division in 1921, setting the stage for cavalry training throughout the 1920s and 1930s. The open areas to the east of the post, although privately owned, served as training grounds for the cavalry troopers. As mechanization began in the Army, Fort Bliss became a proving ground for experimentation. Fort Bliss also hosted cavalry demonstrations, dressage competitions, and other sports events for the troopers.

The rise of fascism in Europe and imperialism in the Far East gained the attention of Fort Bliss leadership. Mechanization and developments in aerial combat would bring change to the post once again and set the stage for a dramatic mission transformation for the post.

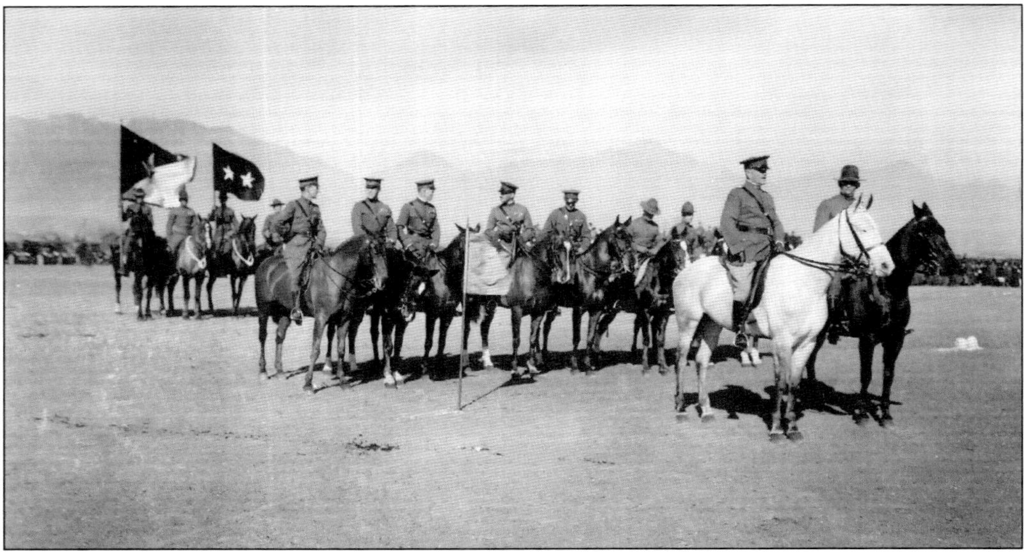

General of the Armies of the United States John J. Pershing reviews the cavalry troops at Fort Bliss in February 1920. This was a national inspection trip, requested by the secretary of war. In El Paso, Pershing saw weapons demonstrations, equestrian exhibitions, an extensive display of weapons and equipment on the main parade field, and a full review of all cavalry troops at Fort Bliss.

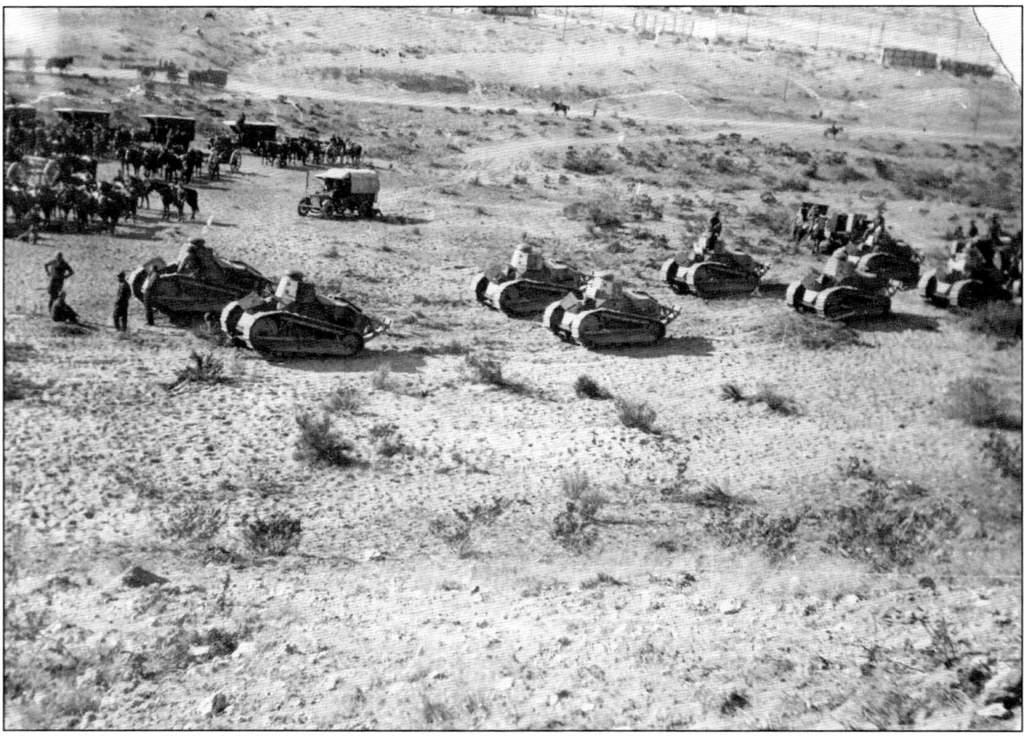

General Pershing's visit included a demonstration of tank combat. The main battle tank was the French Renault FT-17, a very small and primitive vehicle by today's standards. Still, the tanks made a stir among El Pasoans. The demonstration took place at Armstrong Field, looking west toward the mountains, long before Highway 54 was built.

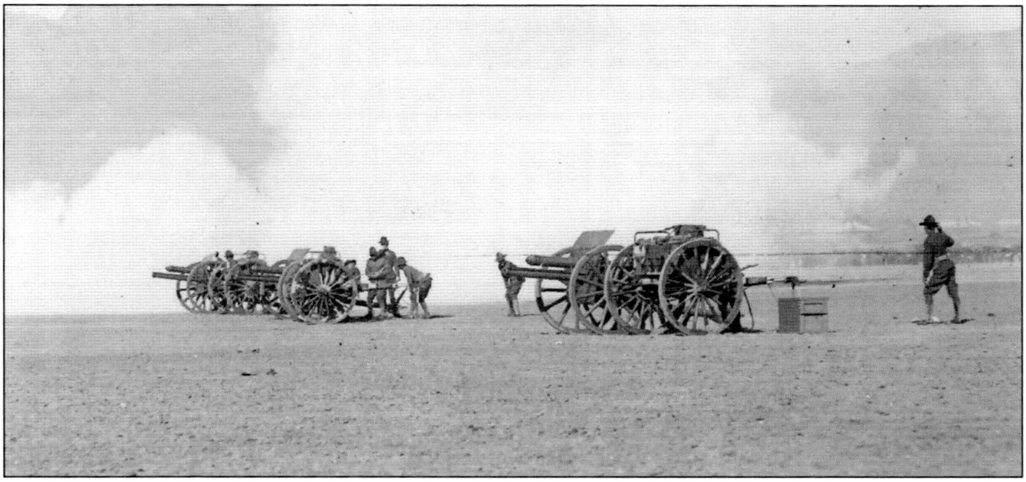

M1917 75-mm field guns fire a salute to General Pershing on Pershing Field Days in 1920. This shot was taken east of the main post, where buildings can just be made out through the smoke. Artillery trained diligently on Fort Bliss, but firing ranges were quite limited at this time. Range land to the east was privately owned, requiring permission from the owners when troops needed to train.

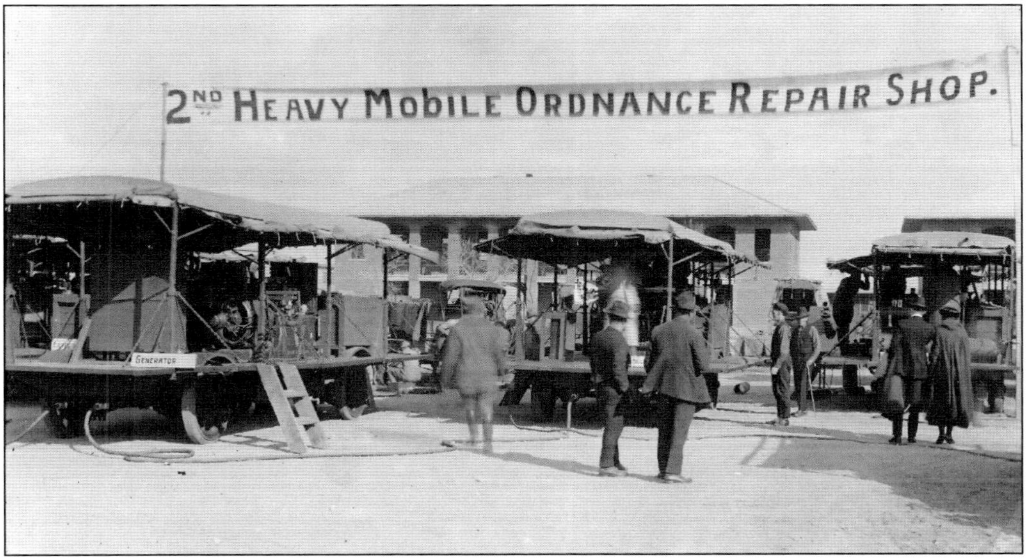

The 2nd Heavy Mobile Ordnance Repair Shop displays its capabilities on the main parade field in 1920. These trailers reflect a higher level of development drawn from World War I experiences. They could generate electricity in the field and make almost any repair to any piece of Army equipment. Note the hard rubber tires on the wheels.

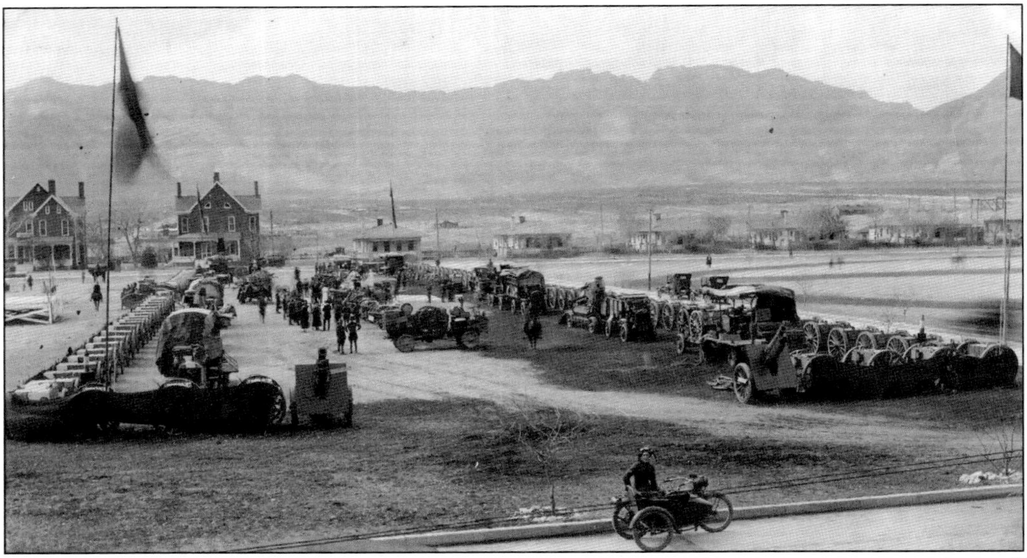

Among other events for General Pershing, the post brought out all its organizations to show off their equipment. All the citizens of El Paso were invited to visit, and many did. Two 155-mm field guns flank the entrance to the exhibition. The trucks, tanks, and field guns are a portent of mechanization. This is about where Building 2 now stands next to Smith Bliss Field.

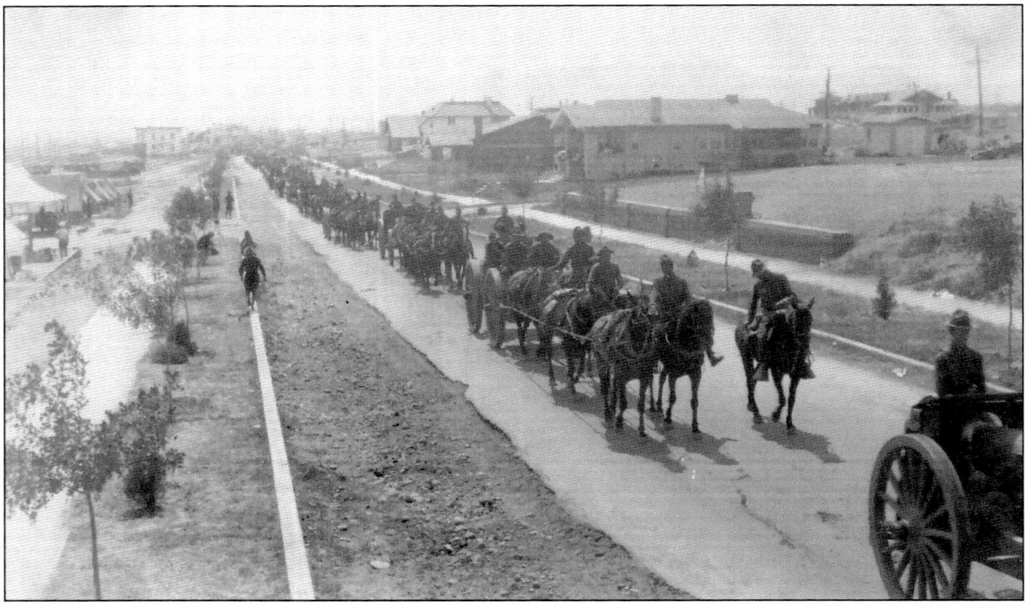

The presence of cavalry brought about numerous parades in and around El Paso. Here, the 82nd Field Artillery passes civilian homes on its way back to Fort Bliss. Artillery cavalcades would always draw crowds of children.

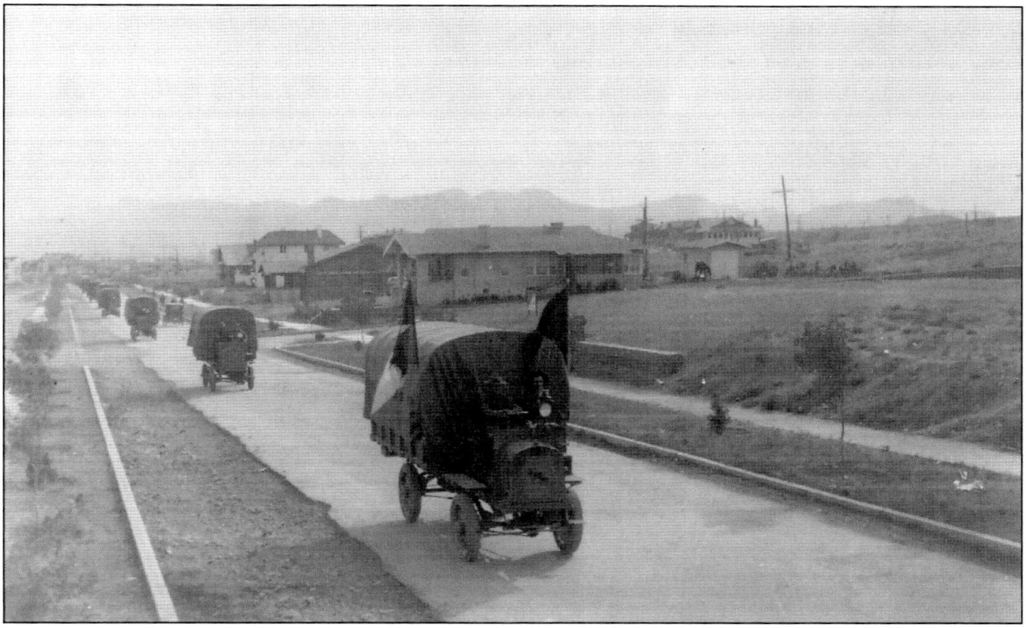

Trucks made by the Four Wheel Drive Company follow the horse artillery north from downtown El Paso. The War Department purchased these for service in the Punitive Expedition without any authorization from the Congress. They served reasonably well, and they were a portent of future Army mechanization. The horse was still dominant at this time, however.

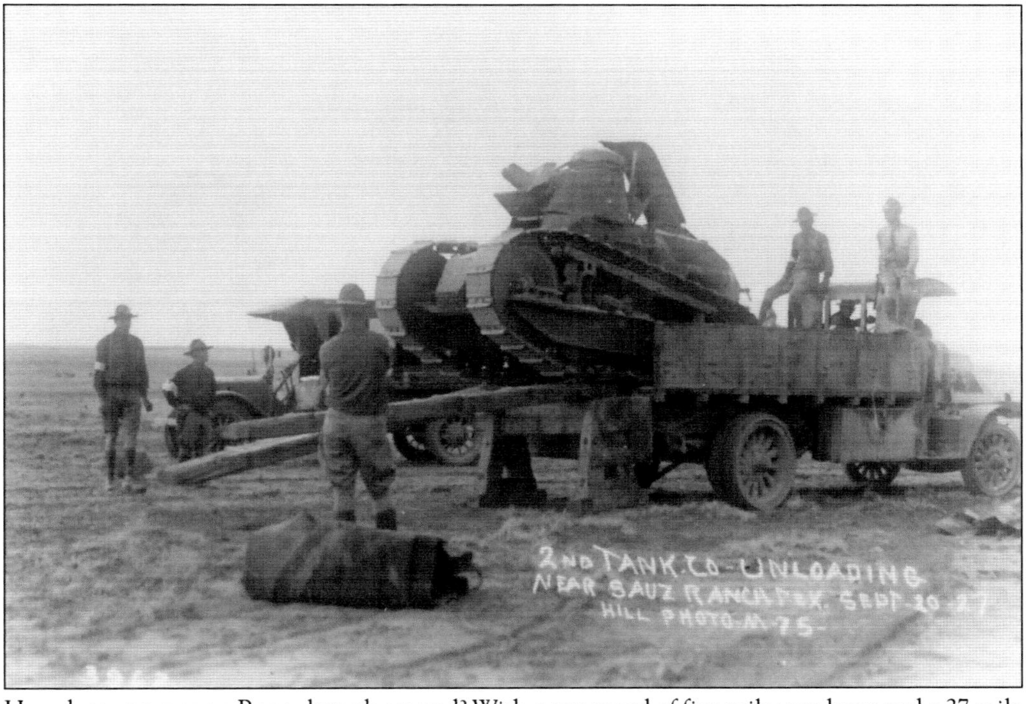

How does one move a Renault tank around? With a top speed of five miles per hour and a 37-mile range, the Renault could not move any real distance. The solution in this case was to load the tank on a truck and haul it to training. Here, troops gingerly drive a tank off a Liberty truck at a training event in 1927.

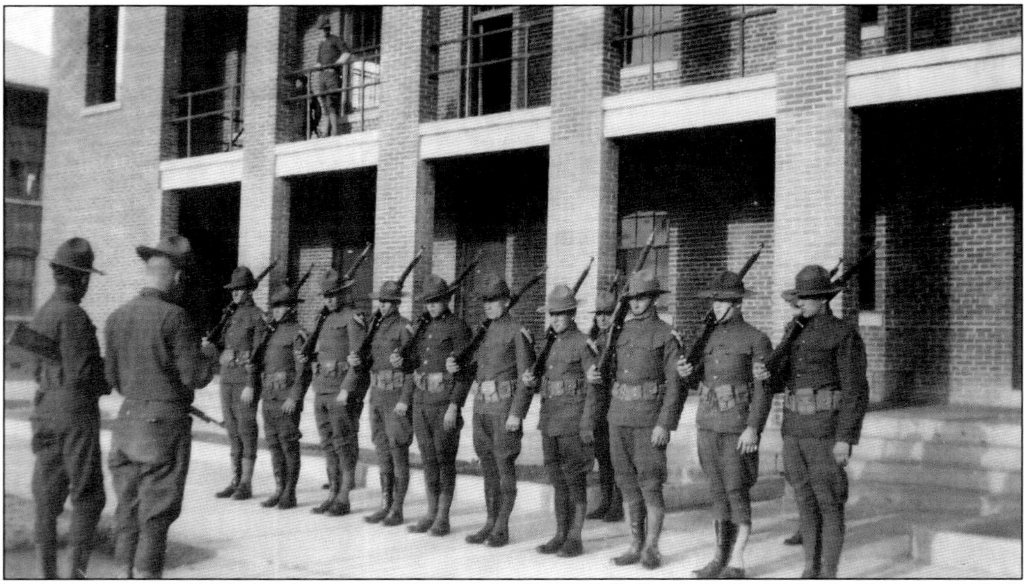

Ten 8th Cavalry troopers are inspected before assuming guard duty. The officer and sergeant of the guard are in front, and one soldier observes from the balcony. This is a cavalry barracks built in 1917 on Pershing Road and still stands at Fort Bliss today.

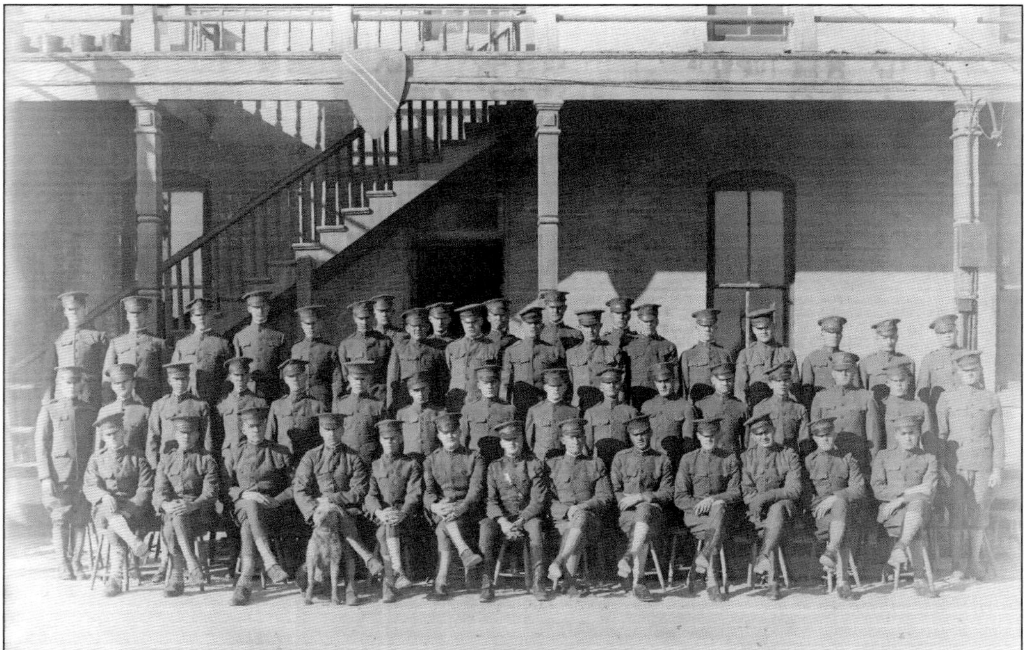

Troops of the 1st Signal Battalion pose for a formal picture in front of their barracks with their mascot, a mixed-breed dog. The weather appears to be as it usually is at Fort Bliss. The picture was likely taken in 1924.

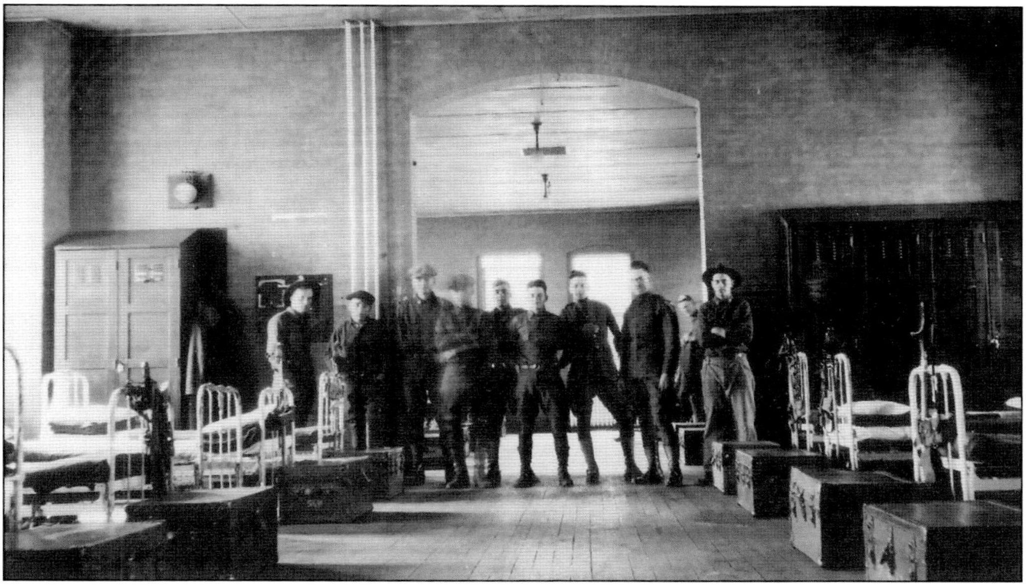
Troops of the 8th Cavalry stand in their barracks on Pershing Road at Fort Bliss. Barracks life was much more austere than today. Troopers had wall-type lockers and footlockers to store their belongings, not much space. Bunks were arranged in rows, with no personal privacy at all.

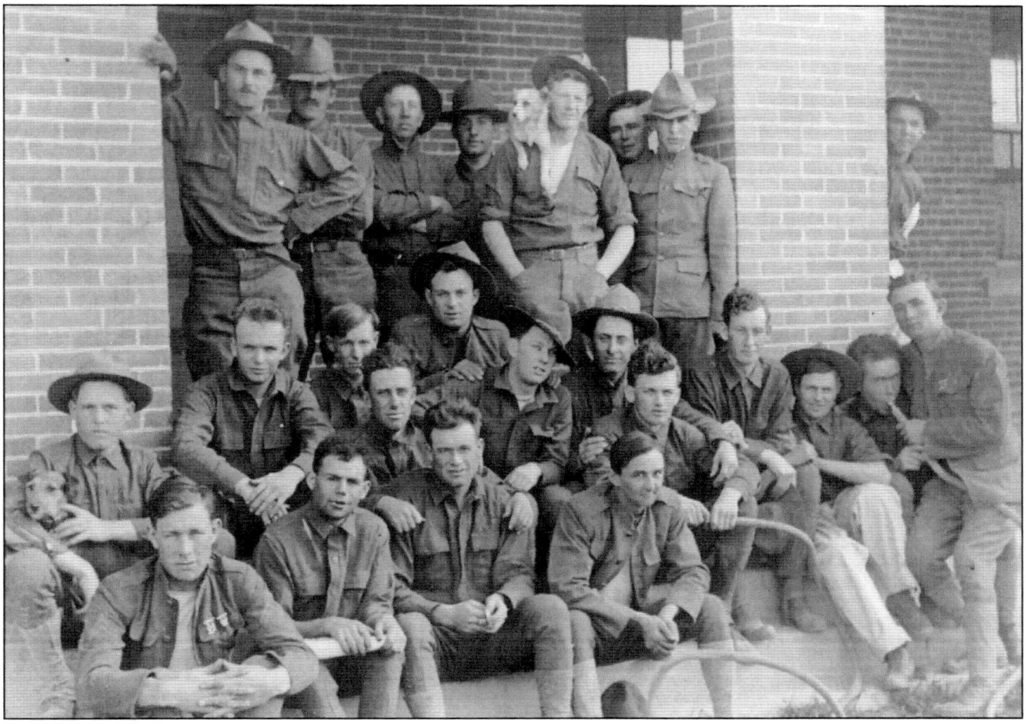
Troops of the 8th Cavalry relax with their dog mascots in front of their barracks in 1922. Mascots were tolerated, not officially approved, as long as they did not become a problem. Some troopers wear their rifle qualification badges, an important part of maintaining training standards. Troopers could earn bonuses by qualifying as marksmen, sharpshooters, or experts with the M1903 Springfield rifle.

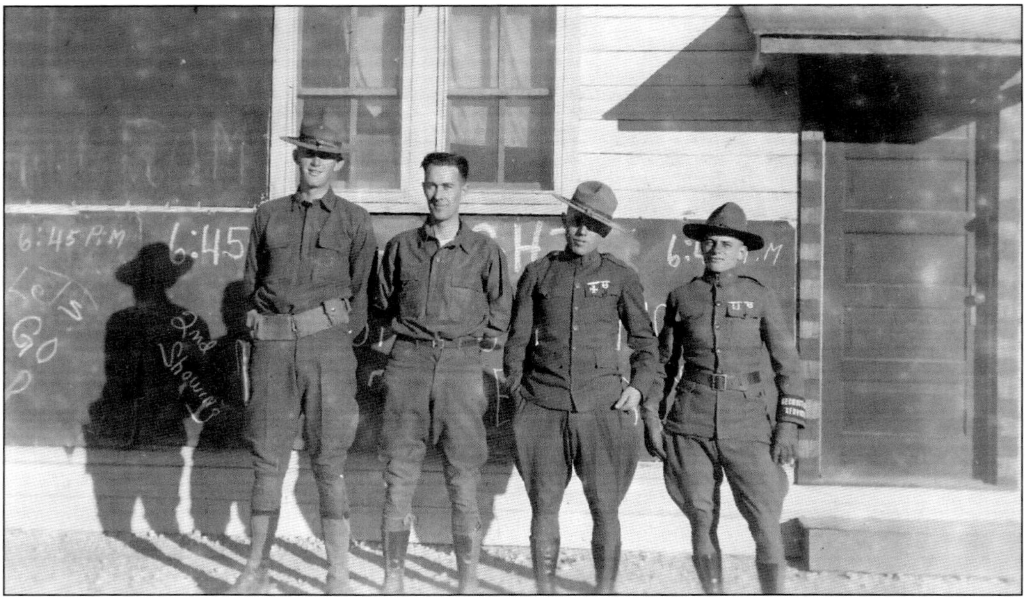

Four troopers stand outside their headquarters at Fort Bliss. The trooper on the right wears a recruiting services brassard, or arm band. Recruiting played as much a part of the Army's mission as it does today. The soldier at far left is likely a duty officer in that he is armed with the regulation M1911 .45-caliber semiautomatic pistol.

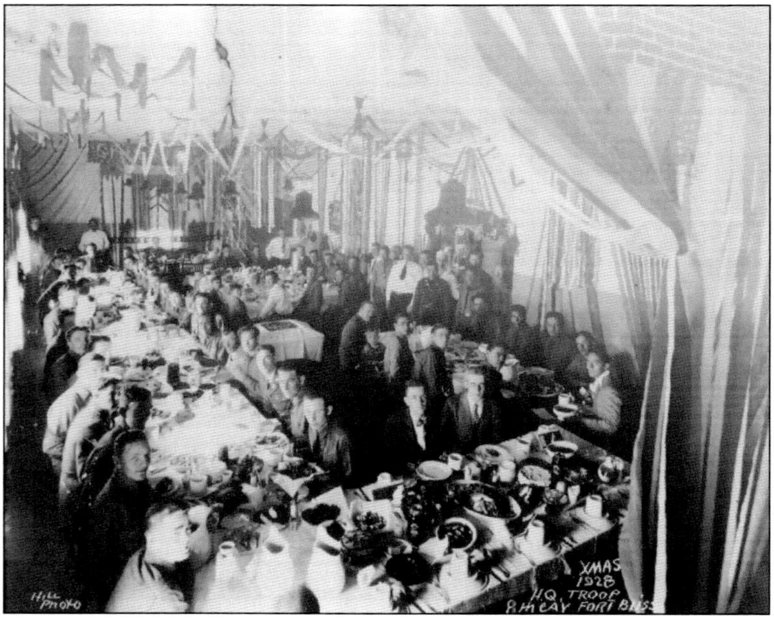

Christmas dinner is served for a troop of the 8th Cavalry Regiment in the 1920s. Cooks served turkeys and hams, freshly cooked vegetables, cakes and pies, and other trimmings. But no alcoholic beverages were served or tolerated. This was during Prohibition, and the War Department took special measures to limit alcohol consumption.

The Army organized the 1st Cavalry Division out of the cavalry regiments at Forts Bliss and Clark in September 1921. This picture shows the division drawn up in a formation to depict the division's shoulder patch. This was part of a calendar produced by an El Paso photography studio. Fort Bliss can be seen in the background.

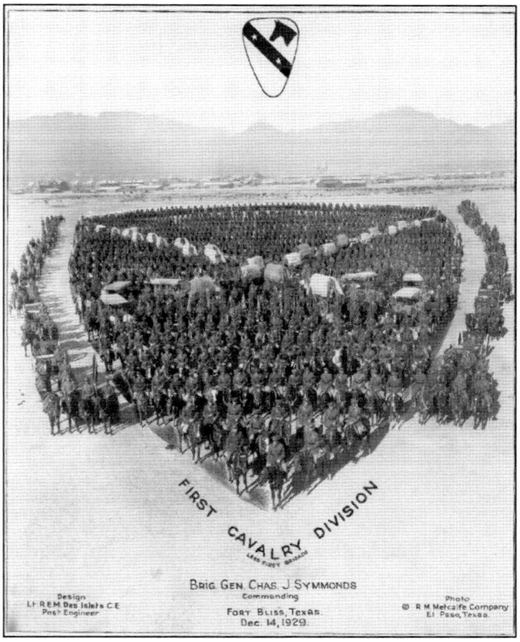

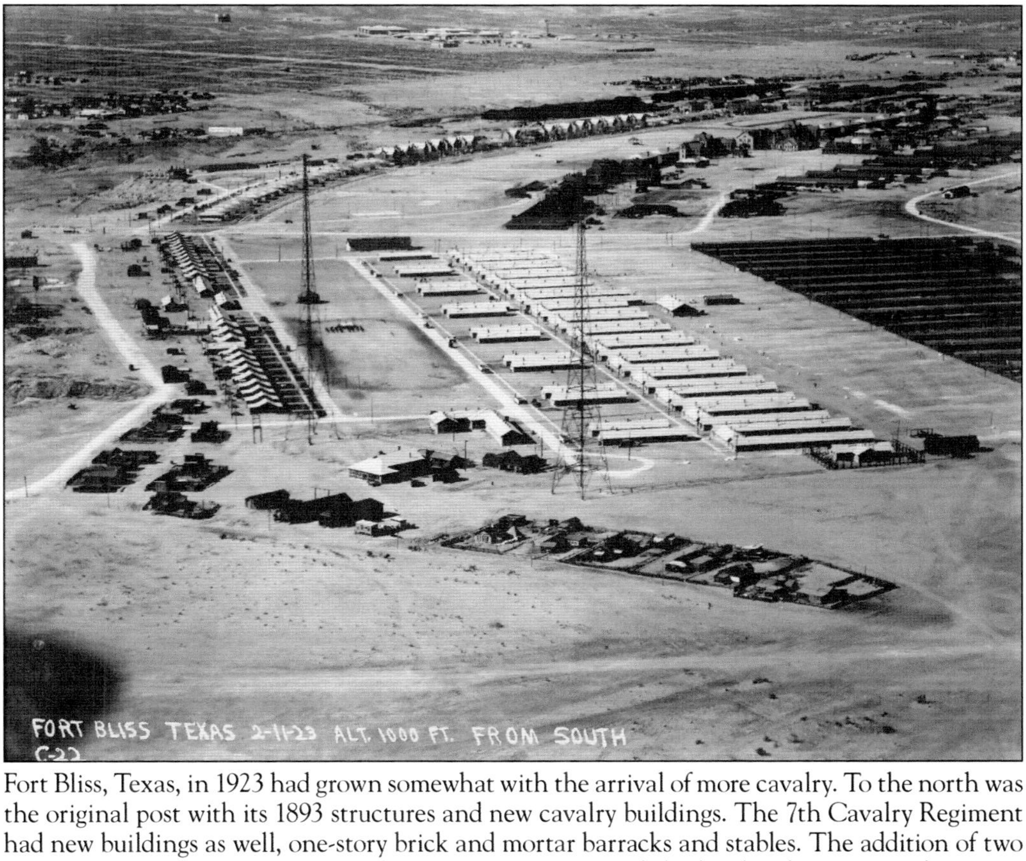

Fort Bliss, Texas, in 1923 had grown somewhat with the arrival of more cavalry. To the north was the original post with its 1893 structures and new cavalry buildings. The 7th Cavalry Regiment had new buildings as well, one-story brick and mortar barracks and stables. The addition of two large radio towers allowed for wireless communication with higher headquarters and units in the field.

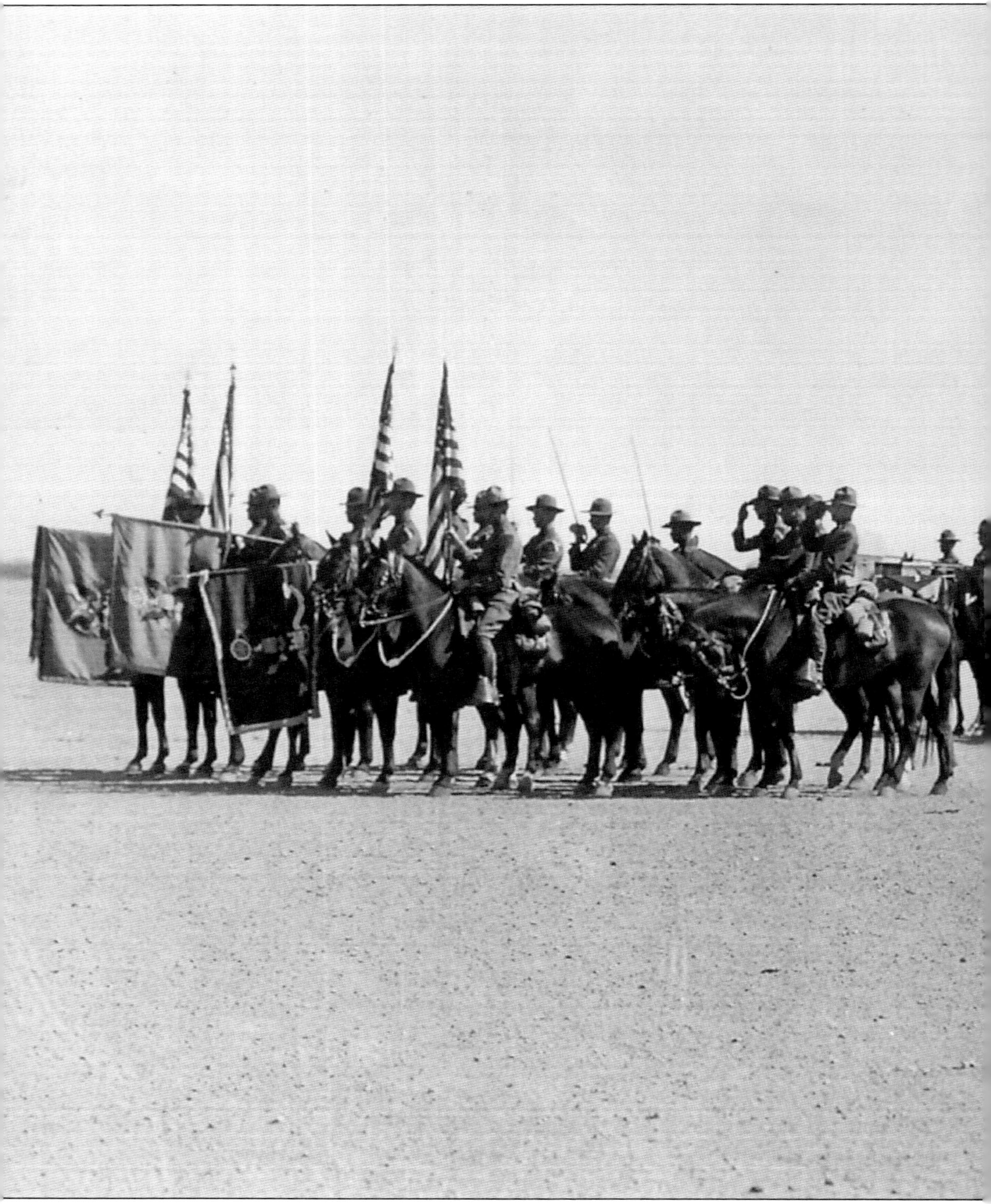

Despite demobilization after World War I, the Army wanted to have divisional units ready for war on short notice. The 1st Cavalry Division was one of those divisions, with its headquarters and the 2nd Brigade at Fort Bliss. Here, the 1st Cavalry Division parades its unit colors for the assistant secretary of the Army in 1922. Visible are the colors of the 5th Cavalry, 7th Cavalry,

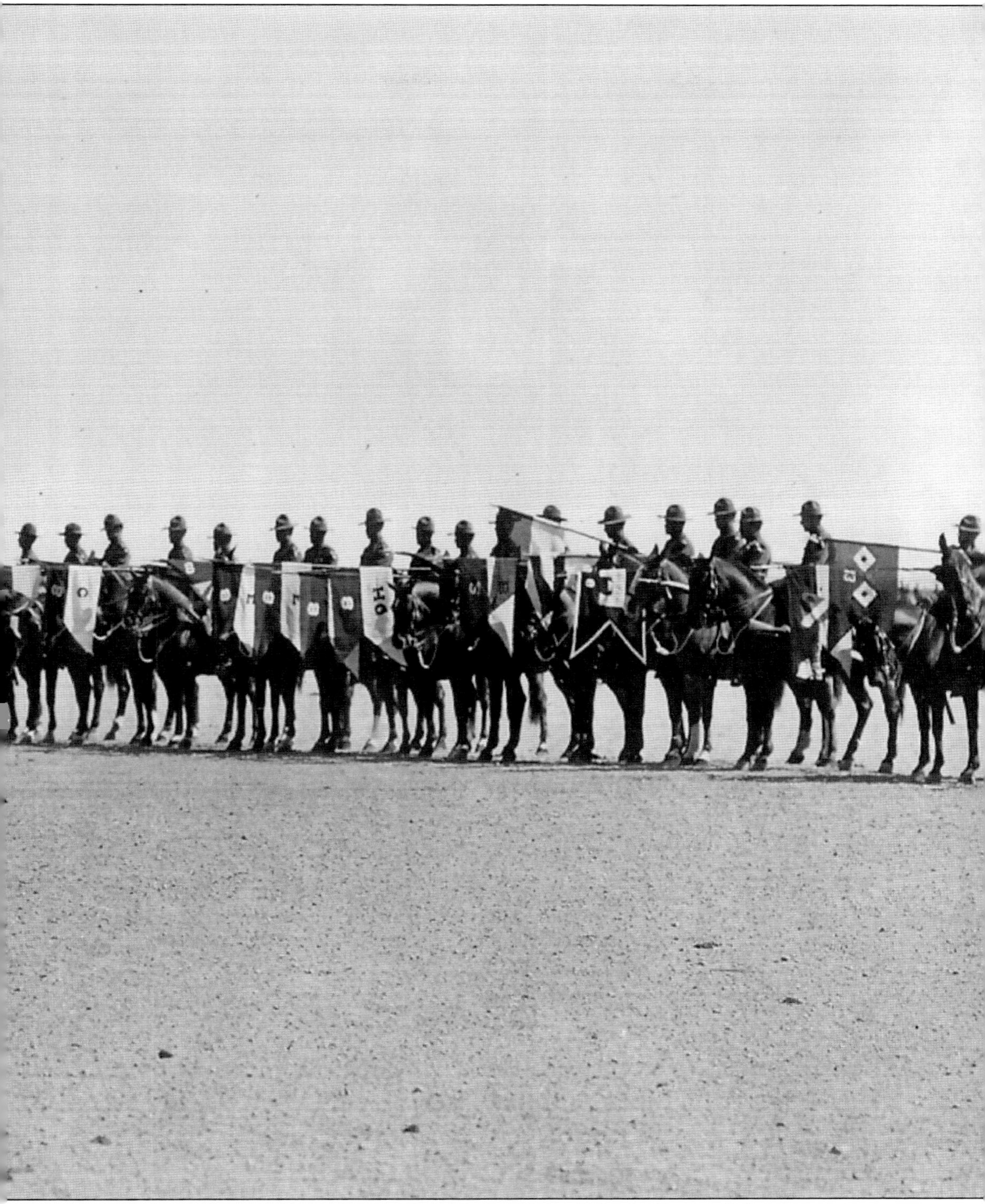

8th Cavalry, 8th Mounted Engineers, and 13th Signal. The soldiers salute with the M1903 Patton saber. Commanded by Medal of Honor winner Maj. Gen. Robert L. Howze, the division was to train for war. Although Fort Bliss was still quite small in terms of training land, Howze arranged to train on ranch land east and north of the post.

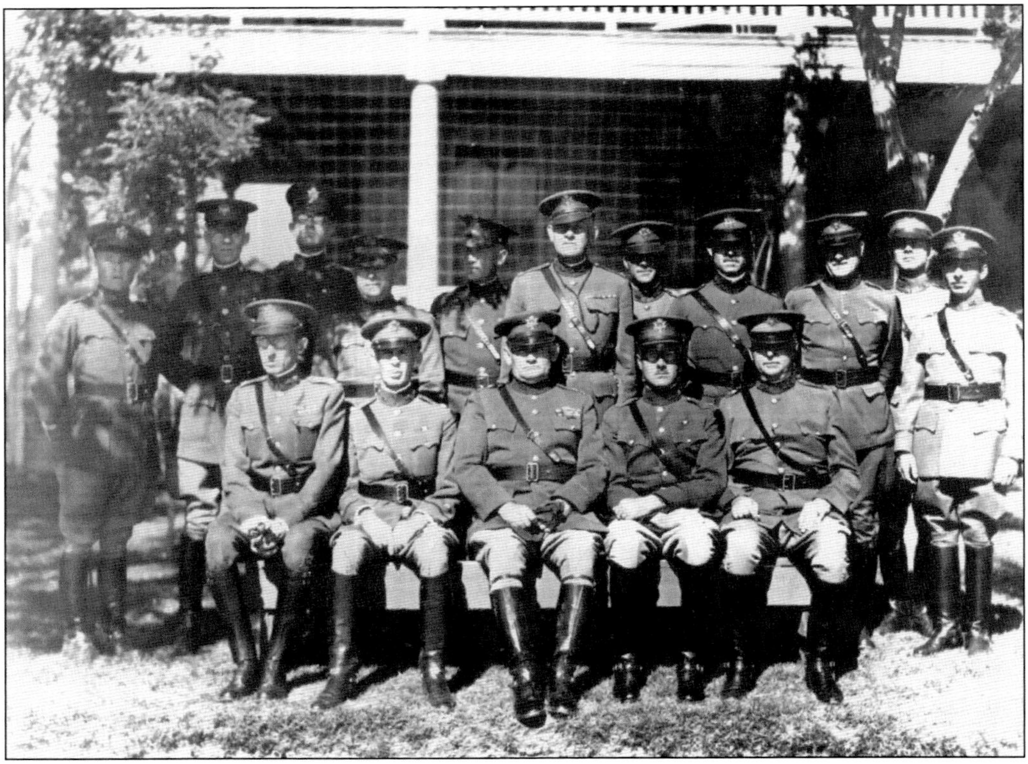

Maj. Gen. Robert Howze sits with his division staff in front of Quarters 1, now named Pershing House. General Howze commanded the 11th Cavalry Regiment during the Punitive Expedition. Howze also served in the Indian Wars, the Spanish-American War, the Philippine Insurrection, and World War I.

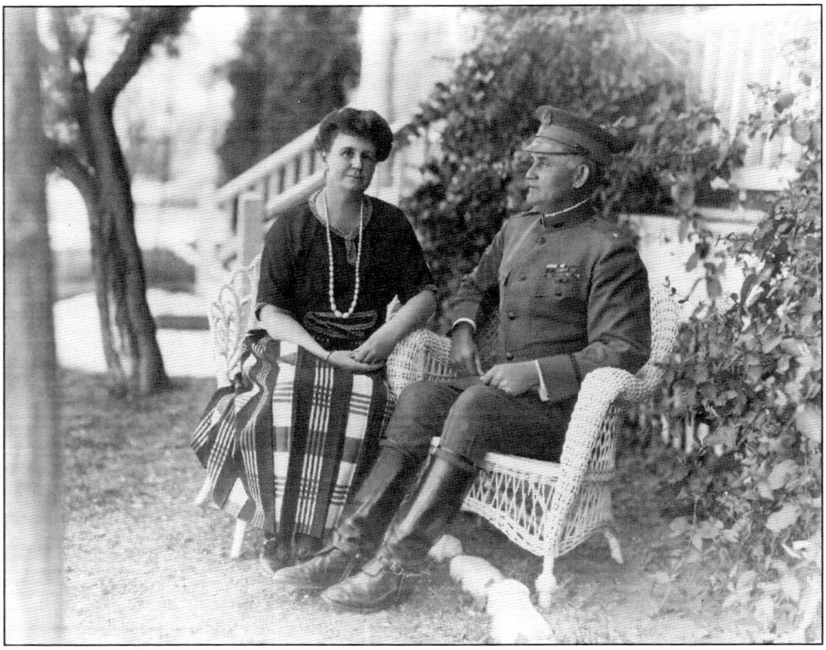

General Howze and his wife, Anne, are seated in front of Quarters 1 in 1921. The house still stands and serves as quarters for senior officers. Generals Scott and Pershing lived in these quarters, as have other distinguished senior officers over the years.

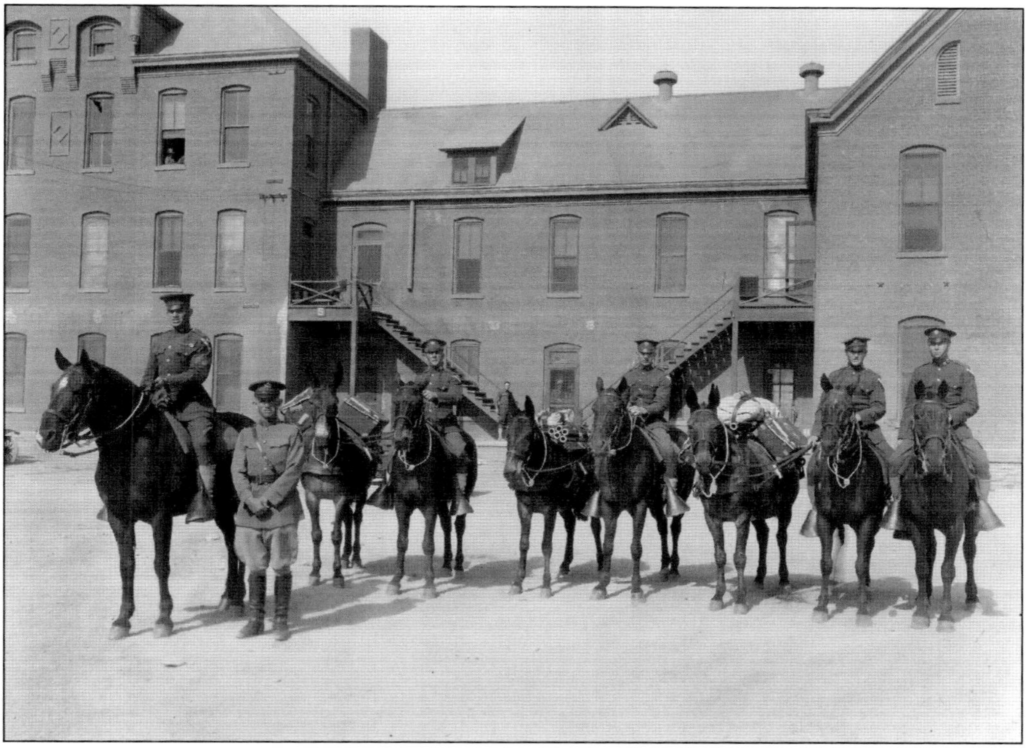

Cavalrymen with loaded pack animals stand behind an original 1893 barracks building at Fort Bliss. Troopers wear the garrison hat rather than the campaign hat. They may be preparing for an inspection or exhibition. They are likely brigade headquarters troopers, carrying tentage and other items for field duty.

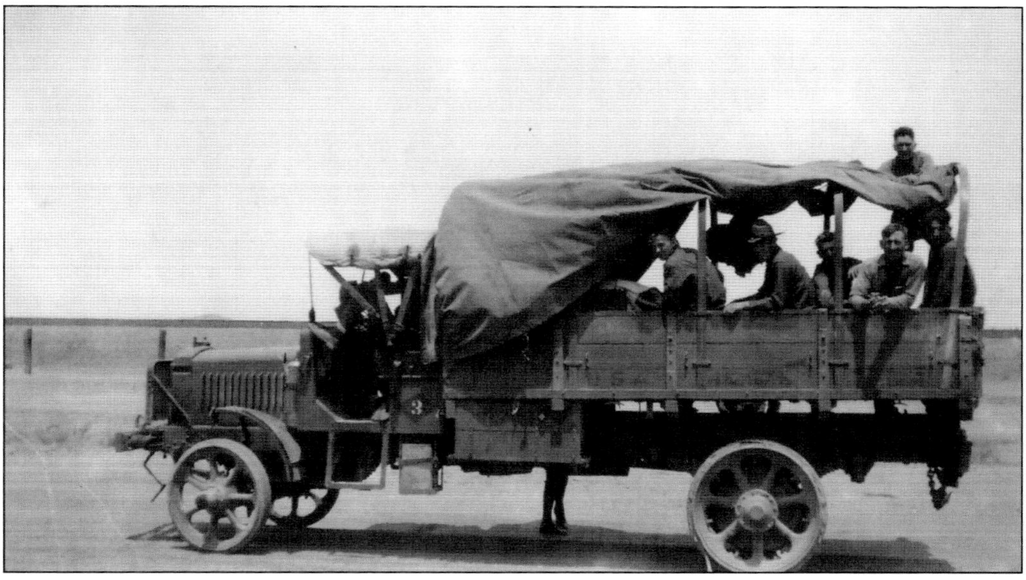

During World War I, the Army adopted the standard Liberty truck. Built by 15 different manufacturers, they were all based on a standard design with interchangeable parts. This was a lesson learned during the Punitive Expedition, where trucks broke down and could not be fixed with parts from other trucks.

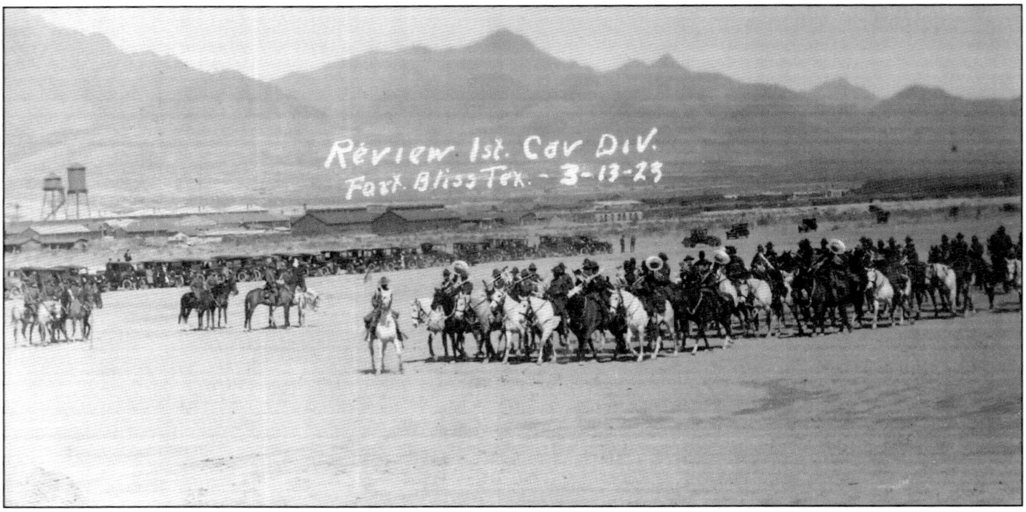

The 1st Cavalry Division Band leads a review east of Fort Bliss in 1923. All band members were mounted and could play well even at a trot. The troopers had to keep their instruments scrupulously clean given the dusty conditions. At left rear are two water towers; only the water tower at right remains today.

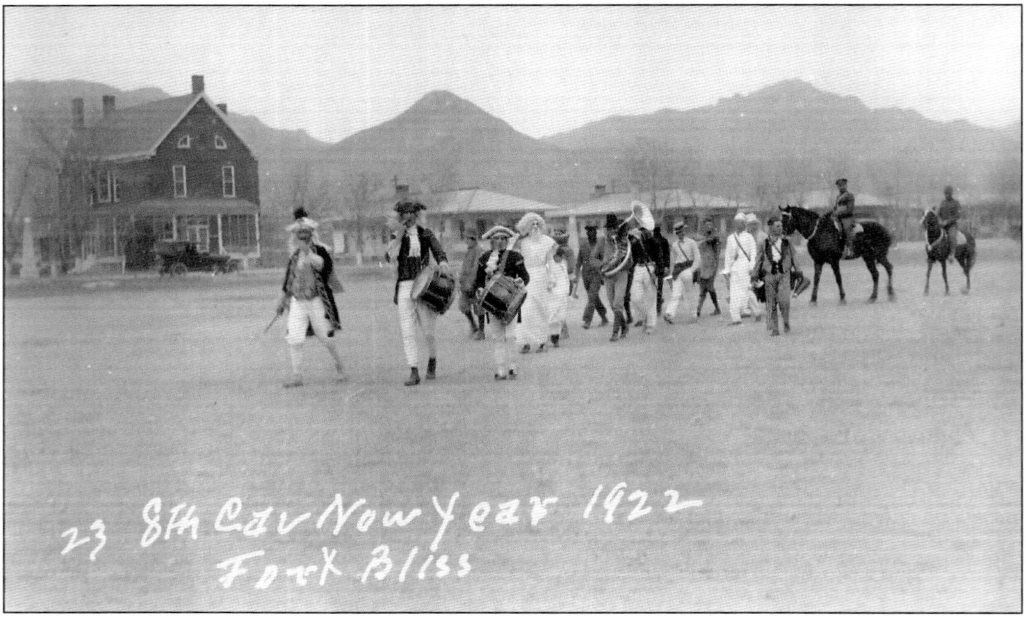

The 8th Cavalry troopers celebrate New Year's Day in 1922 on the main parade field. They wear improvised costumes and parade around between their barracks and the officers' quarters. There may have been some alcoholic beverages involved, even though Prohibition was certainly in effect.

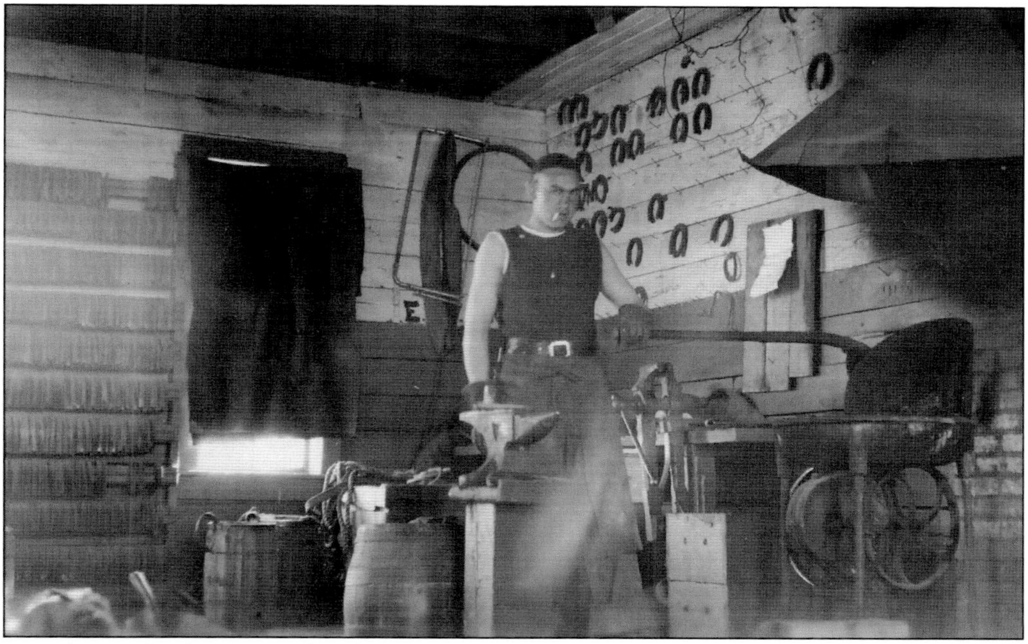

An 8th Cavalry farrier stands at his forge. Every regiment had qualified farriers, and each troop had a soldier trained to shoe horses. Most went through the post's horseshoeing school. Every soldier carried several horseshoes in their saddle kits, available if a horse threw a shoe on campaign. Today, soldiers carry extra track blocks on their tracked vehicles.

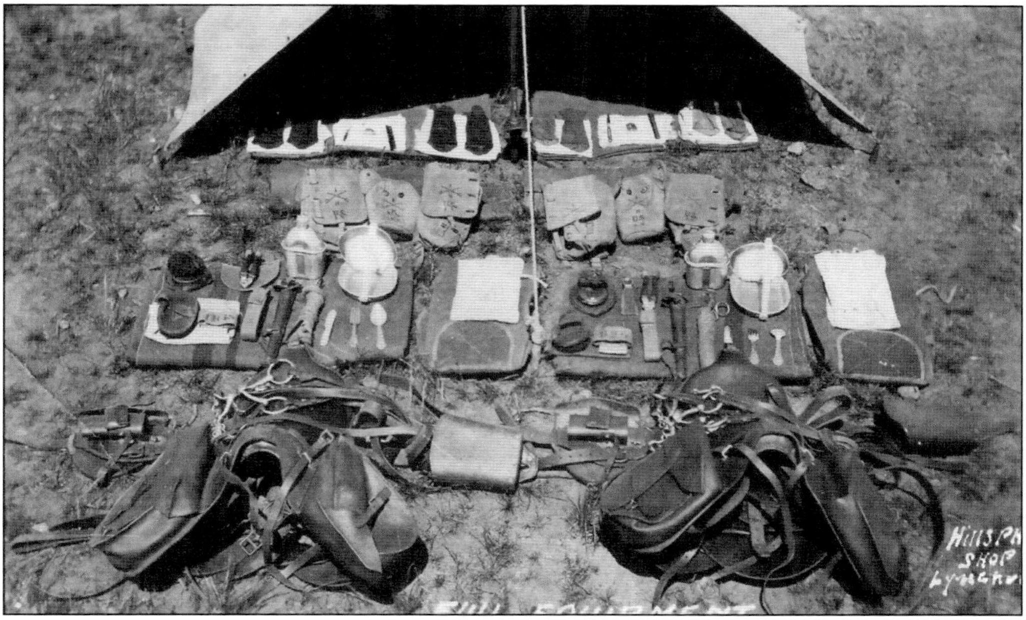

Before going to the field for training, troopers endured a full layout of equipment for inspection. Canteen covers indicate that these troopers were in the 11th Cavalry. Everything was inspected, including saddles, blankets, and even shaving equipment and toothbrushes. On returning from the field, the troops underwent an inspection again.

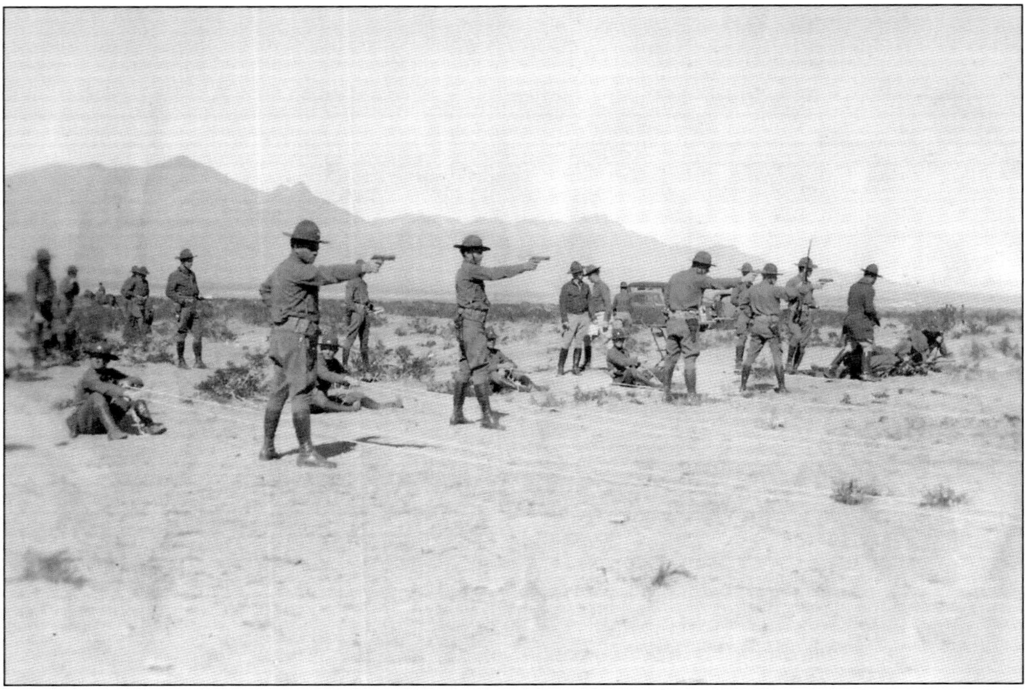

The .45-caliber automatic pistol replaced the saber in mounted combat after its adoption in 1911. First, soldiers learned to fire the pistols on a range, then they learned to do it while mounted at a gallop. These troopers practice on a range east of the post, toward what is now the El Paso International Airport. In those days, the land was empty.

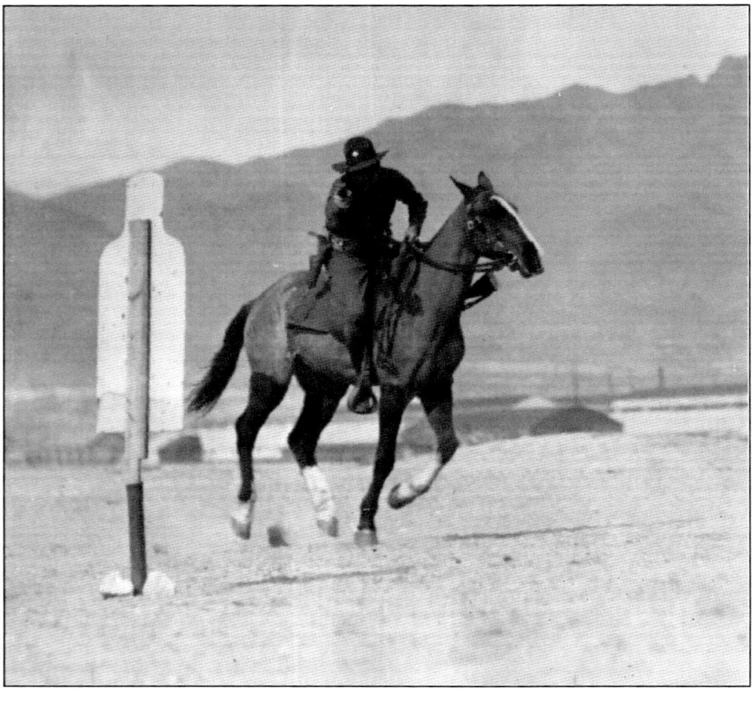

Mounted horsemen were at a disadvantage without a firearm. The M1911 .45-caliber automatic pistol provided needed firepower and served better than a saber; however, the range of this pistol was quite short. Riders needed to gallop close to their targets before firing. Here, a trooper fires at a silhouette target east of the barracks.

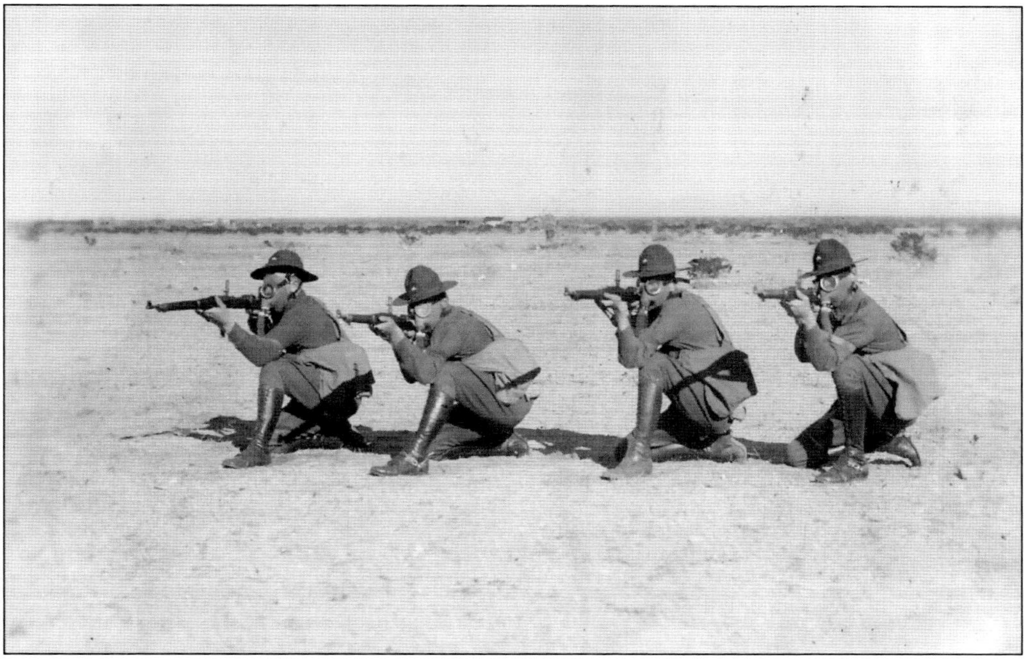

With the advent of chemical warfare in World War I, troops now had to train to do all their combat tasks with gas masks on. This included marksmanship training, as these troopers demonstrate here. Horses were fitted with masks as well, which may or may not have provided protection in a gas attack. Nevertheless, this was part of a cavalry soldier's training.

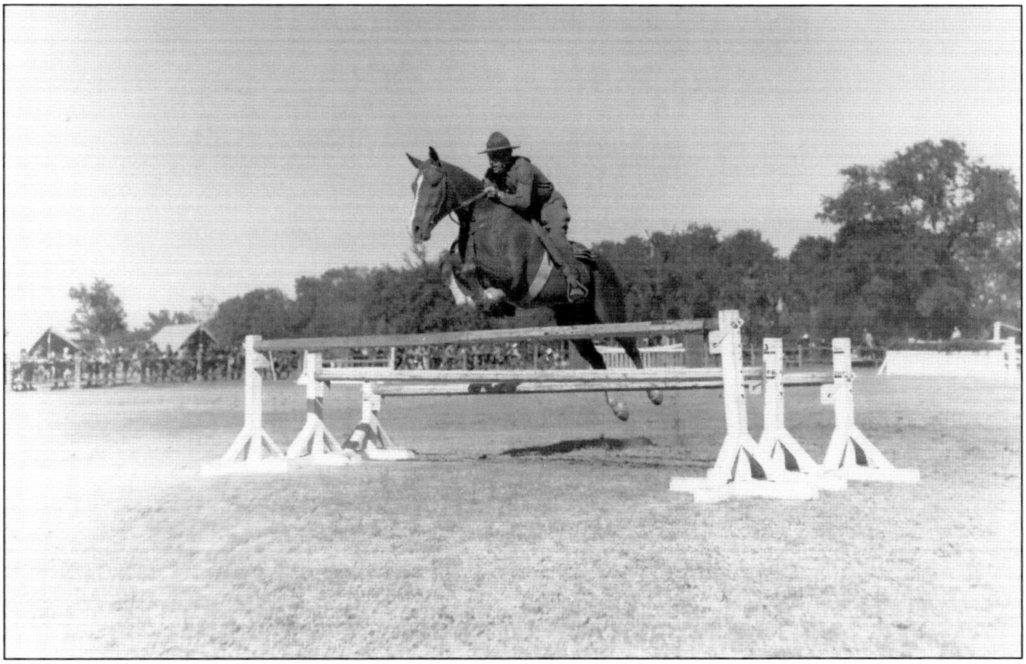

The most critical part of training, though, was equitation, or horsemanship. Soldiers trained with their mounts on the main post and in field training. The soldier had to be one with the horse. Here, a soldier jumps his mount over an obstacle at Noel Parade Field. These obstacles were on the main parade field until the 1990s, well after horse cavalry disappeared.

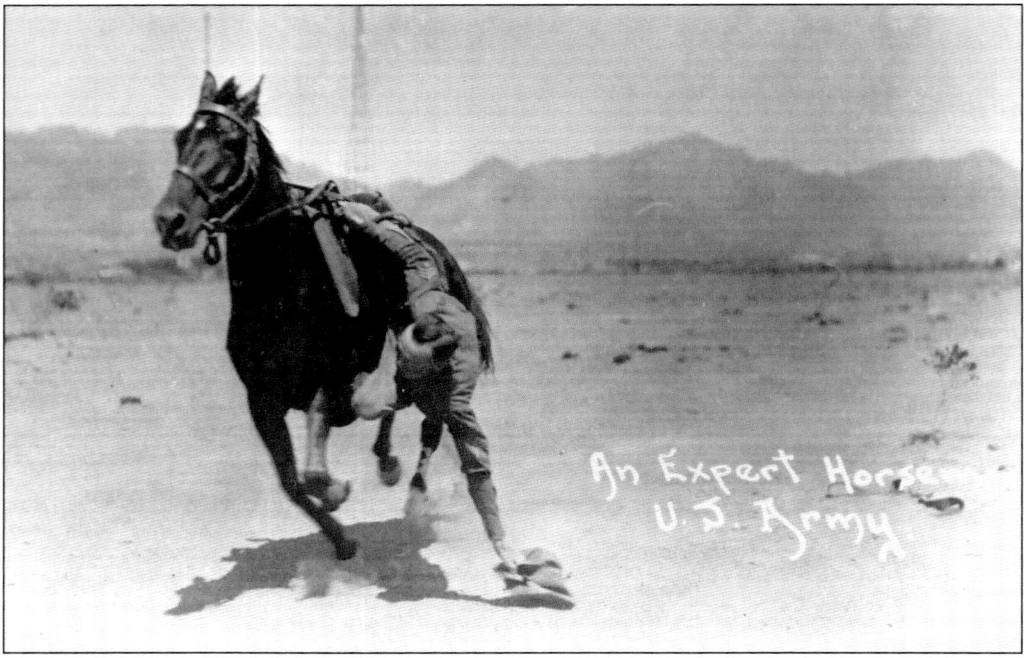

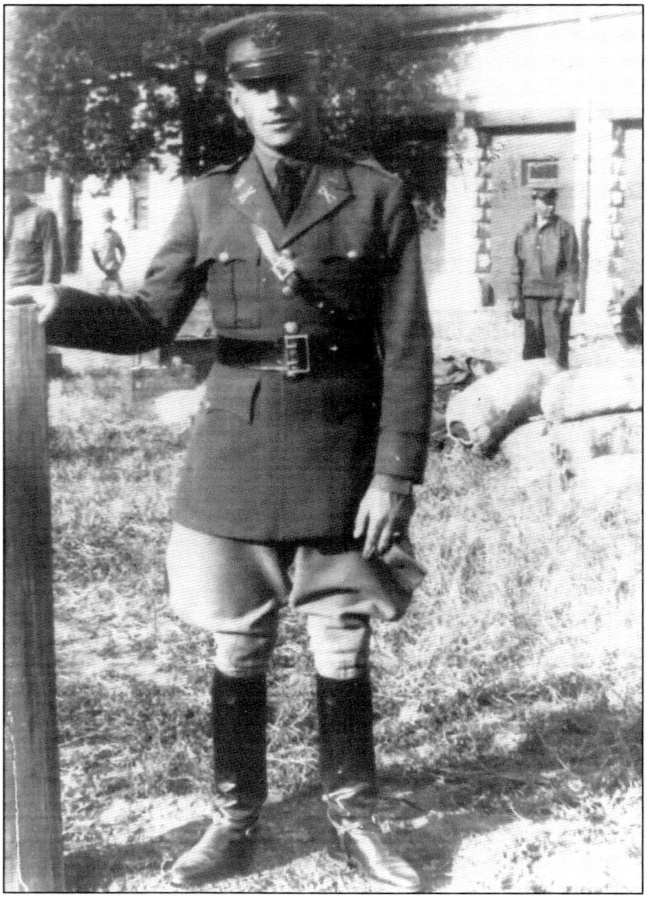

The expert horseman could do amazing feats with his horse, including picking up his campaign hat at full gallop. This was not without risk. If the horse stumbled, the rider could expect to be thrown into the dust. Sometimes, the trooper got right back on and tried again. Other times, he would be seeing the regimental surgeon at the hospital.

The main parade field next to the cavalry barracks was named Noel Parade Field in 1927, after Lt. Paul Alpheus Noel. Lieutenant Noel was killed while playing polo on this field, demonstrating that it was a dangerous contact sport.

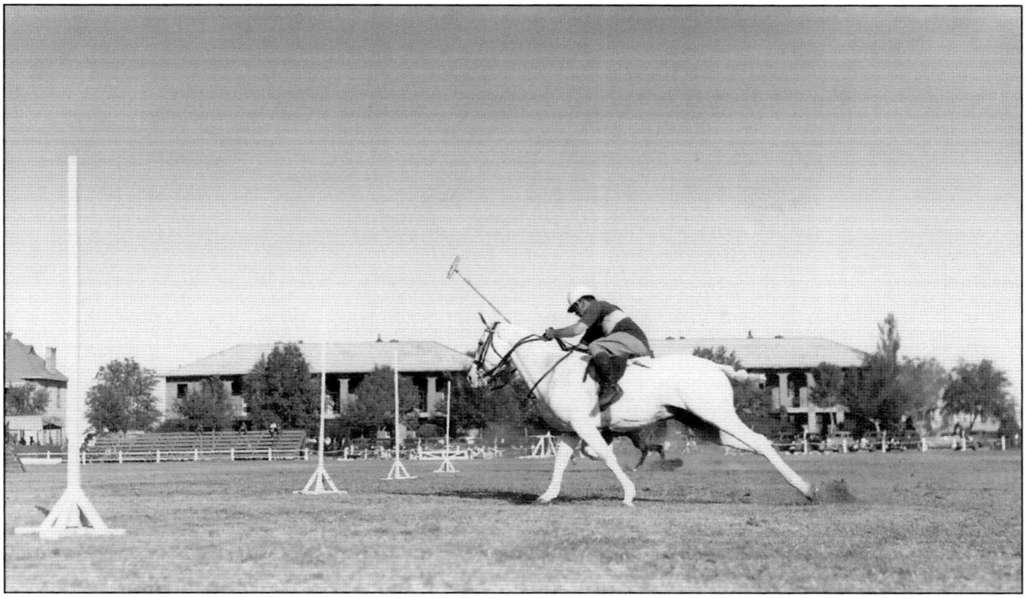

But polo played a strong part in training the cavalry soldier. Officers played against each other, and units fielded polo teams that competed against each other as units. Competition produced proficiency at horsemanship, expected to pay dividends in combat operations against a live enemy. Smith Bliss Field was a favorite exhibition area during the cavalry period at Fort Bliss.

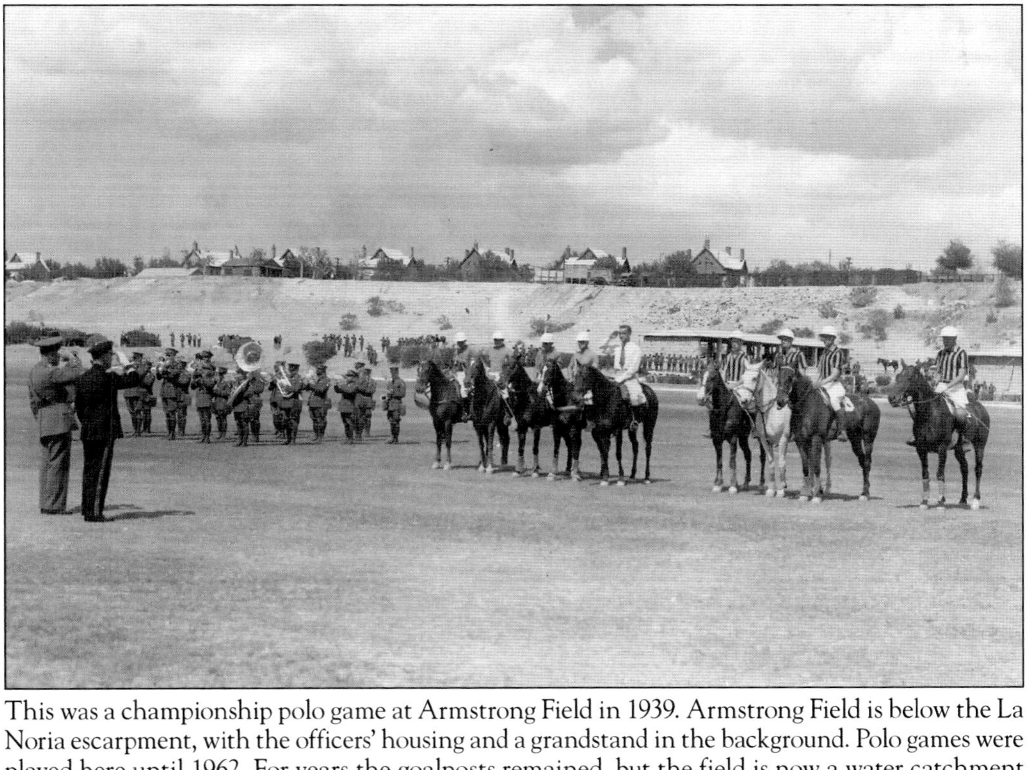

This was a championship polo game at Armstrong Field in 1939. Armstrong Field is below the La Noria escarpment, with the officers' housing and a grandstand in the background. Polo games were played here until 1962. For years the goalposts remained, but the field is now a water catchment basin for the city of El Paso.

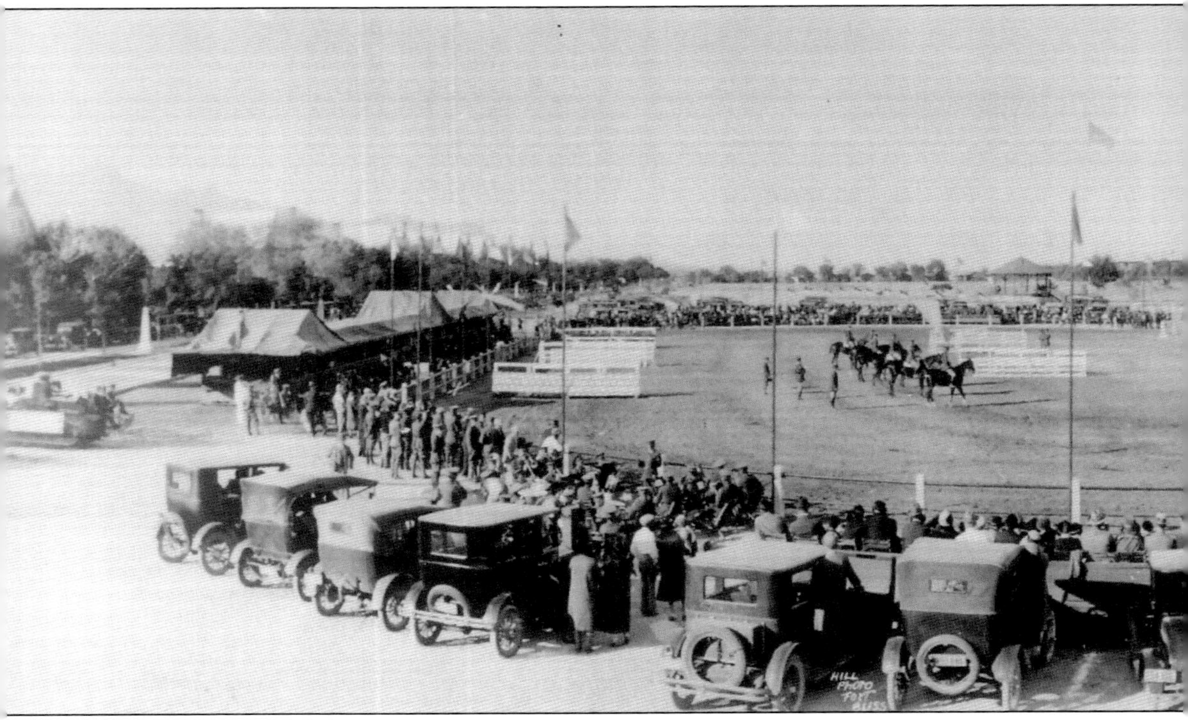

Troopers participate in a major equestrian exhibition in Howze Stadium at Fort Bliss in the 1920s. The stadium is surrounded by automobiles, indicating that El Pasoans were on hand to view the activities. A tank sits at far left. Officers' housing is on the left, and the barracks are on the right. The troops have put up sun shelters for the reviewing stand to provide shade in the late

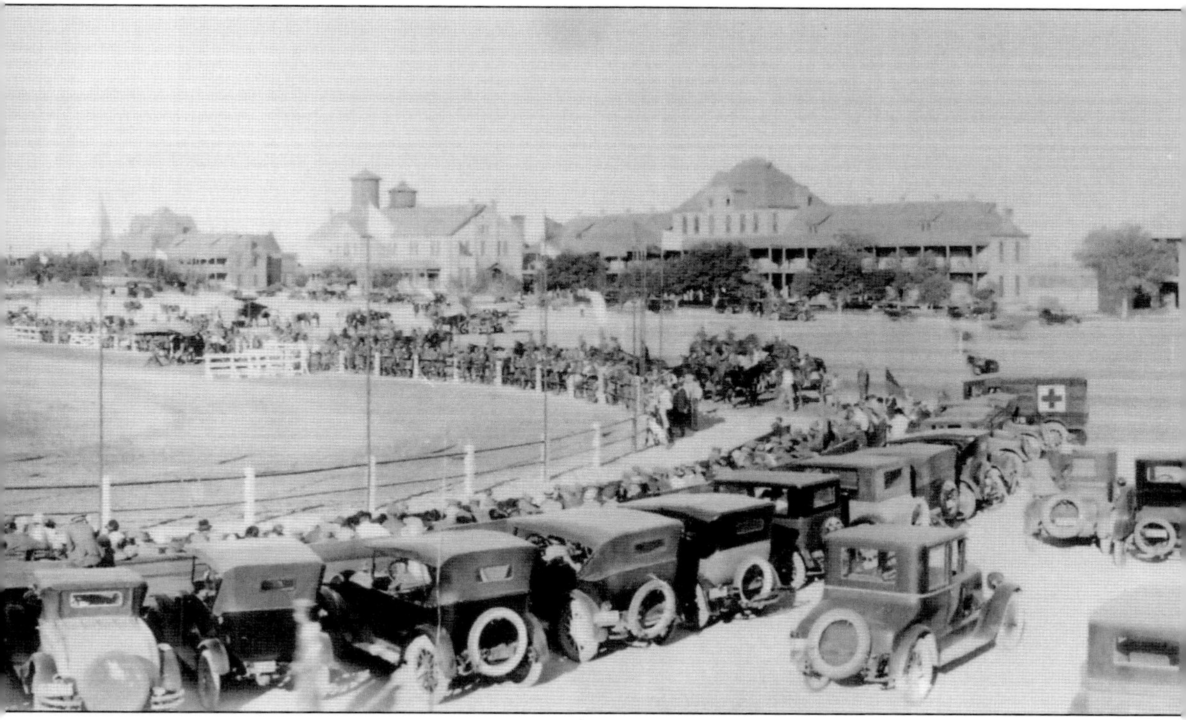

afternoon. The horse jumps indicate that this included a dressage exhibition. On hand at right is an obligatory ambulance, necessary for any trooper who was injured. The pennants around the field indicate that this is a relatively calm day. The spectators are dressed for what appears to be a cool day in early spring.

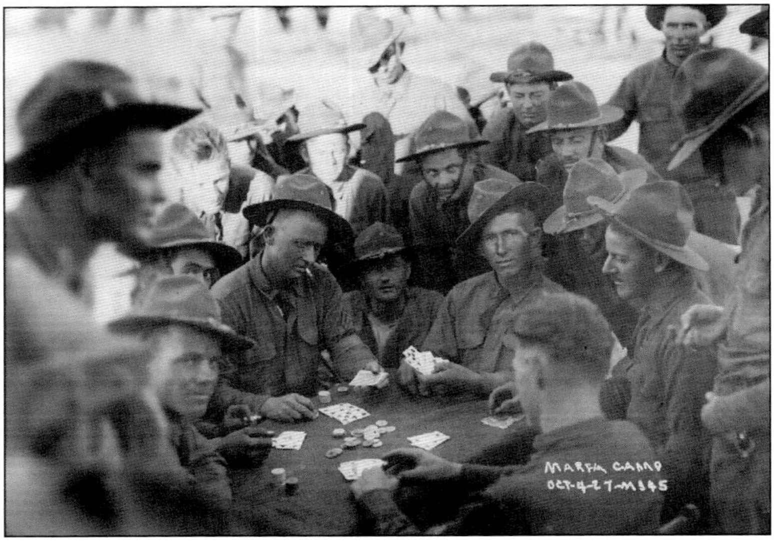

Polo and other training took up only so much time. Troopers amused themselves in different ways. Cavalry soldiers of the 1st Signal Battalion pass the time in field by playing cards. They appear to be using poker chips, but a $1 bill is visible. This was in camp near Marfa, Texas, during a 1927 training exercise.

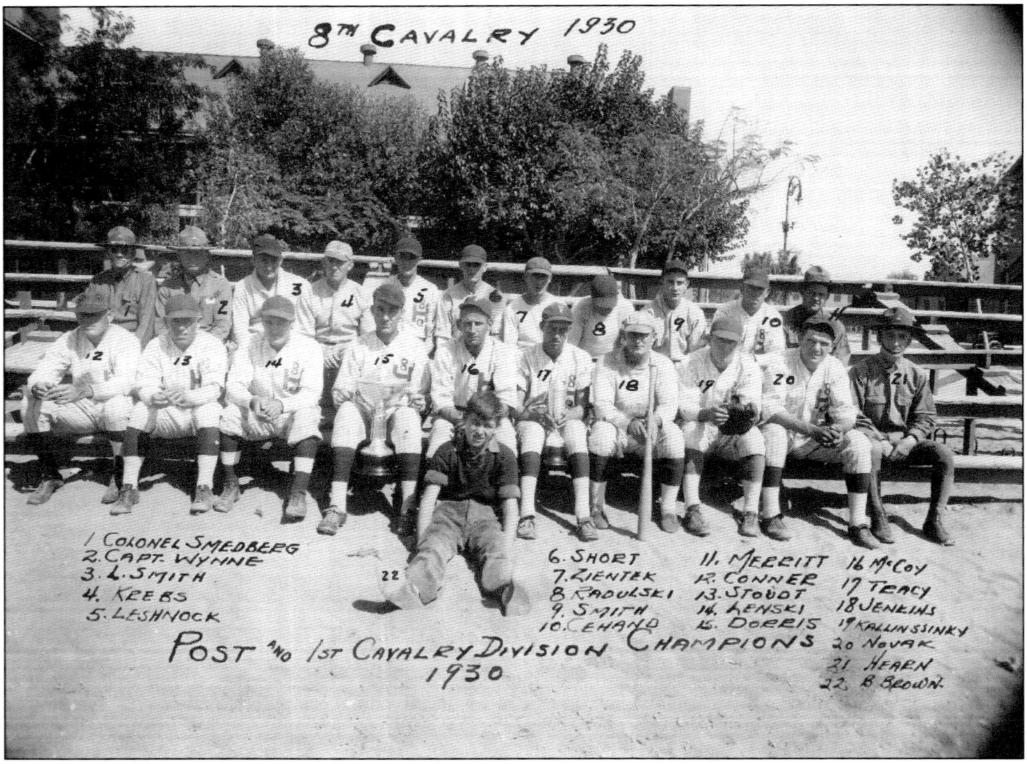

Sports events were very popular among commanders and troops, with major competitions between units. The 22 members of the 8th Cavalry baseball championship team pose in 1930. The regimental commander and the troop commander pose with them, basking in their success. The picture was taken in front of their barracks, either Building 11 or 113.

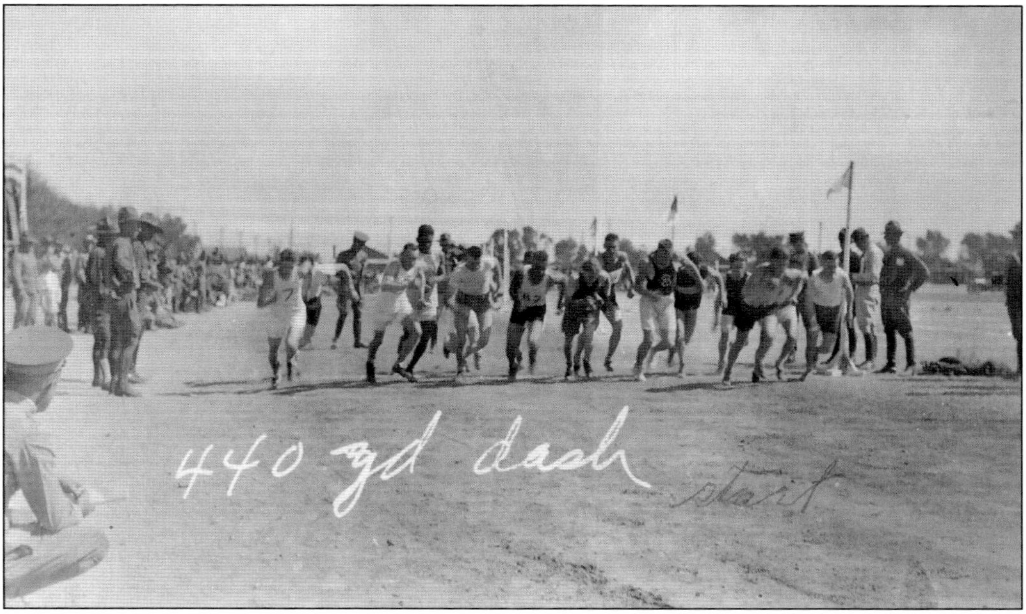

The start line for a 440-yard dash event sits on Smith Bliss Field. Troopers participated in track and field events, competing against other troops and squadrons. Here, they start the race, overseen by their officers and NCOs serving as timekeepers.

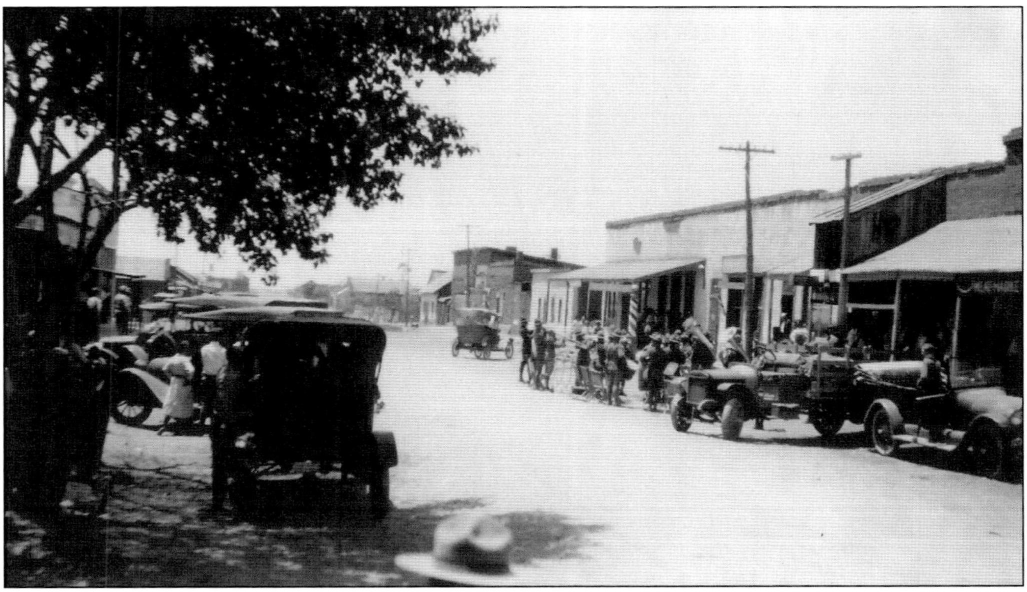

But for all the wholesome activities promoted by their leaders, troopers were also lured by unwholesome pursuits as well. The small community of Lynchville, established to the northwest of the post, had saloons and other sordid attractions that drew soldiers and their pay. Here, a military band plays in front of a barbershop at the approximate intersection of Dyer Street and Fred Wilson Road today.

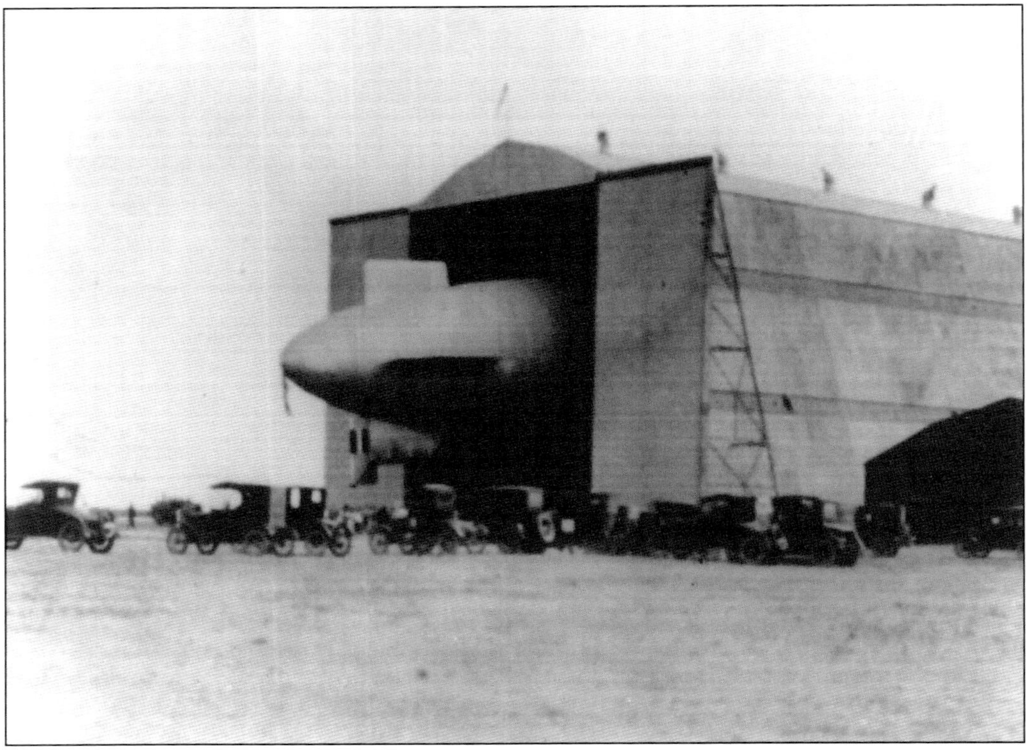

Biggs Army Airfield began as the Fort Bliss Flying Field in 1921. The airfield was named after Lt. James B. Biggs, who died in a plane crash in France in 1918. This was a small airfield in the 1920s and 1930s, hosting patrol planes along the border. This hangar held military blimps, existing until it was torn down in 1966.

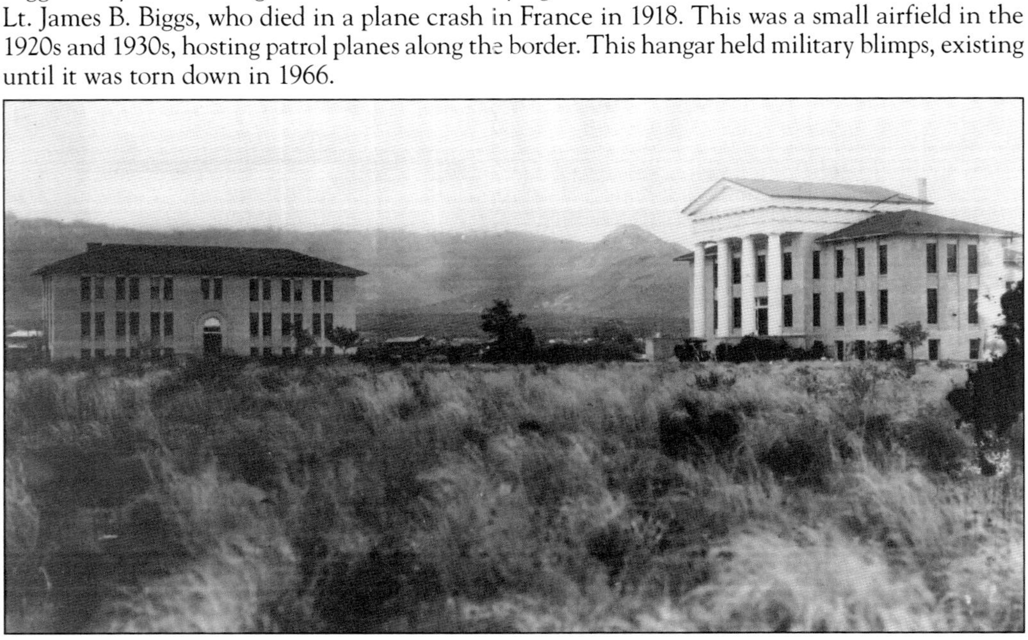

East of Fort Bliss was the old El Paso Military Institute and School of Mines and Metallurgy. These buildings later became the headquarters of the Fort Bliss Flying Field until Biggs Army Airfield was established. The School of Mines moved to what is now the campus of the University of Texas at El Paso. These buildings no longer exist. Buildings east of Freedom Crossing stand in their place today.

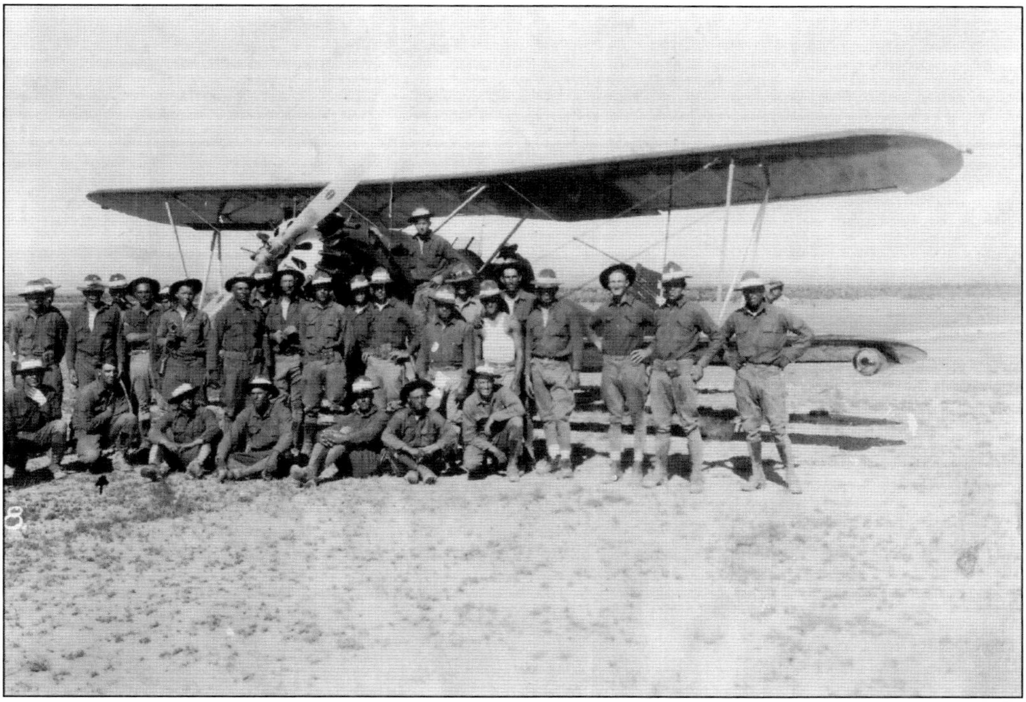

Cavalry troopers stand next to an airplane, apparently a Boeing F2-B that was used as an observation plane. Troopers are on maneuvers and wear white tape on their hats to identify them as part of the "White Army," opposing the other half of the division that was the "Blue Army." The airplane was one of five attached to the cavalry on an operation near Berino, New Mexico, in 1930.

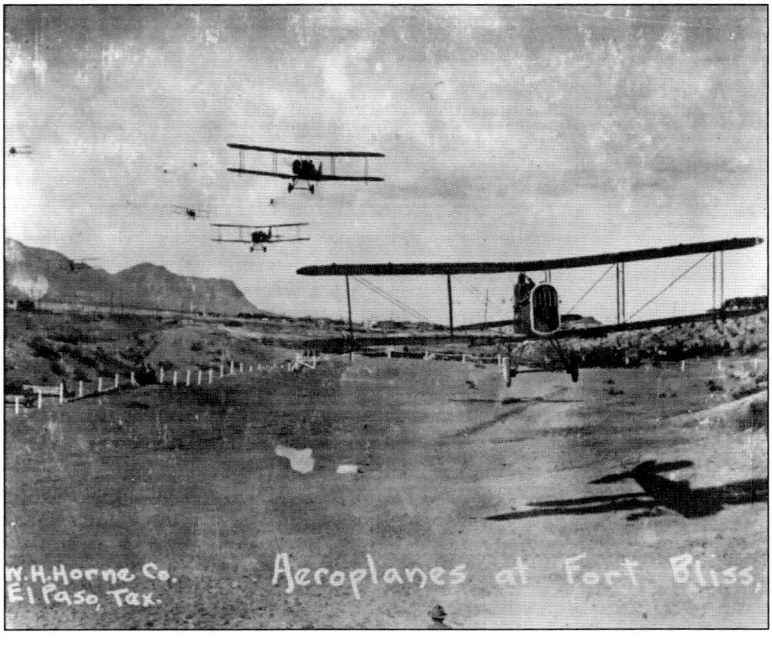

A flight of DeHavilland DH-4 airplanes fly low over Armstrong Field during an exhibition in 1917. The DH-4 was underpowered and tended to be unstable, and pilots referred to them as "flying coffins." At the end of World War I, they were replaced by more powerful airplanes.

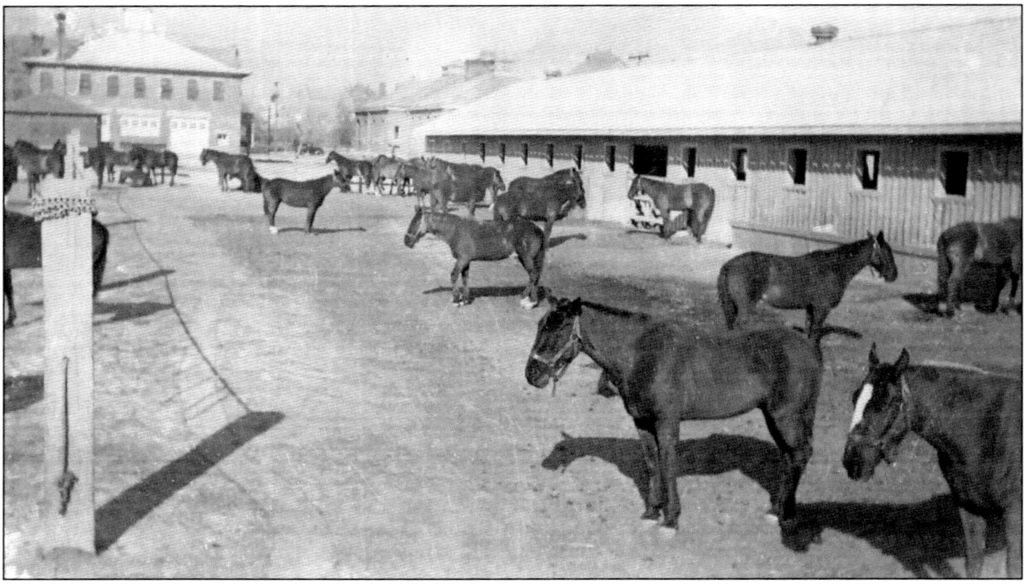

Despite the progress of motorization and flight, the main forms of transport in the cavalry days were horses and mules. In the 1920s, the horse corrals were just east of the barracks on Pershing Road. This image shows a corral next to the fire station. The station still stands, but the horse corral is long gone. This made mounts quickly available, and the training and range areas were a short ride to the east.

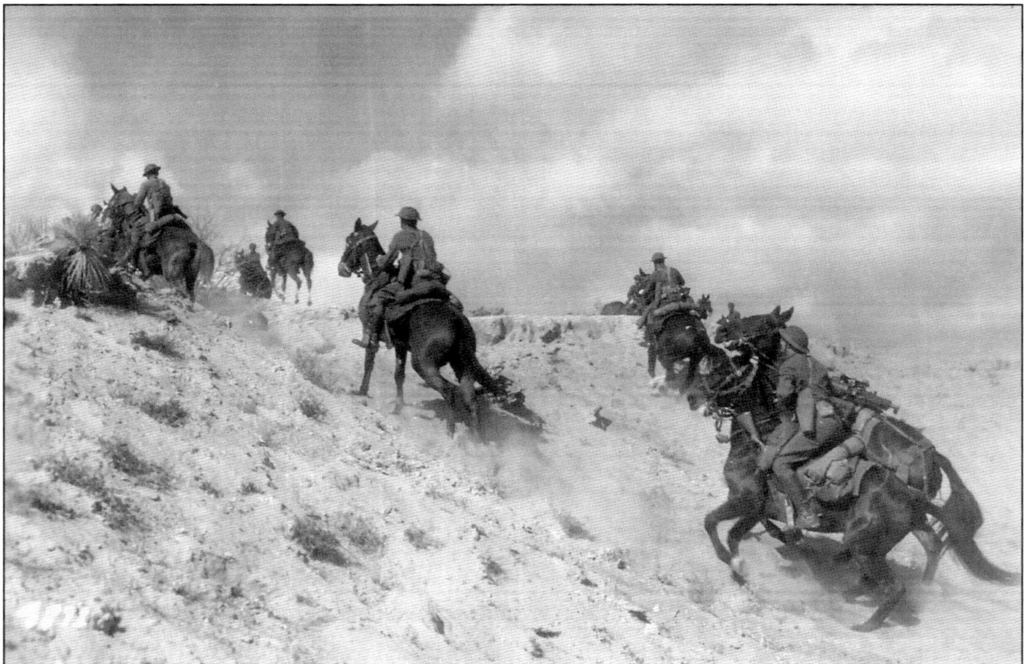

A horse cavalry machine gun team is in action, charging uphill. The team carries the Browning M1917 .30-caliber machine gun, standard issue through World War II. This gun was water-cooled and rather heavy, but the horses appear to be carrying them well.

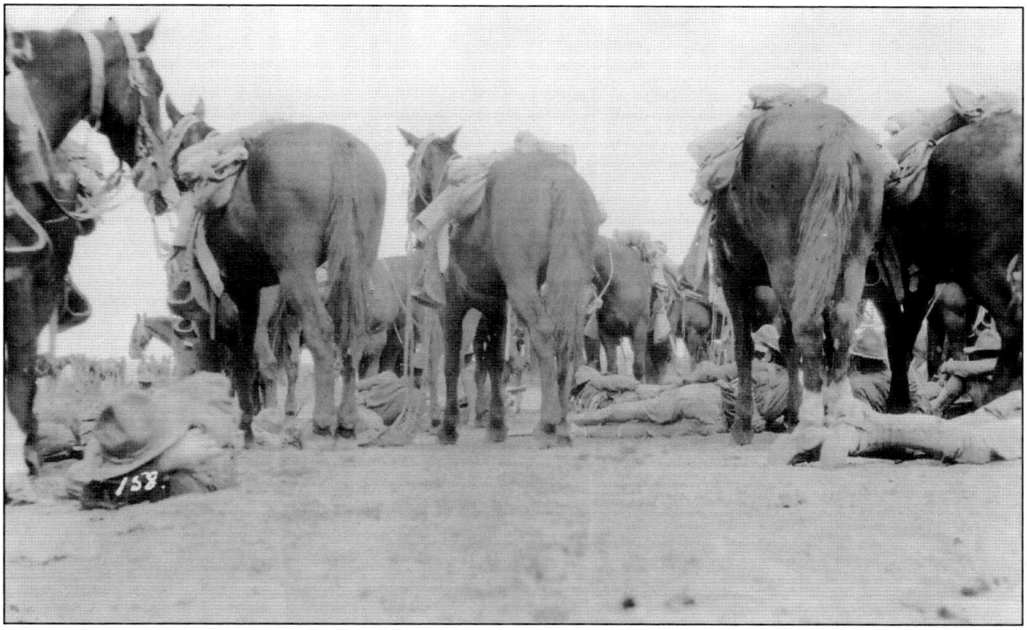

Troopers rest underneath their mounts in the field. This seems rather dangerous—what if the horses stepped on the troopers, or worse? But after a hard training ride, troopers rested where they could and kept their mounts immediately available.

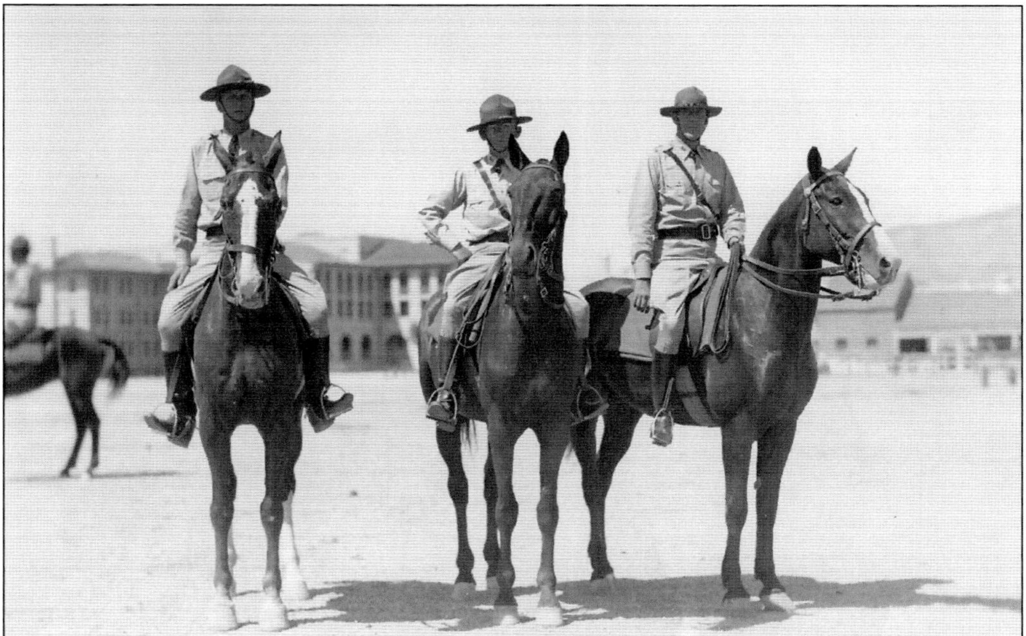

Troopers of the 7th Cavalry Regiment mounted on Noel Parade Field. These troops stand in front of their brand-new three-story barracks, built in 1938. These barracks housed the 1st Cavalry Division, a better housing arrangement than it previously had. These barracks were built as aa Works Progress Administration (WPA) project. Still standing, the barracks are now used as administrative space.

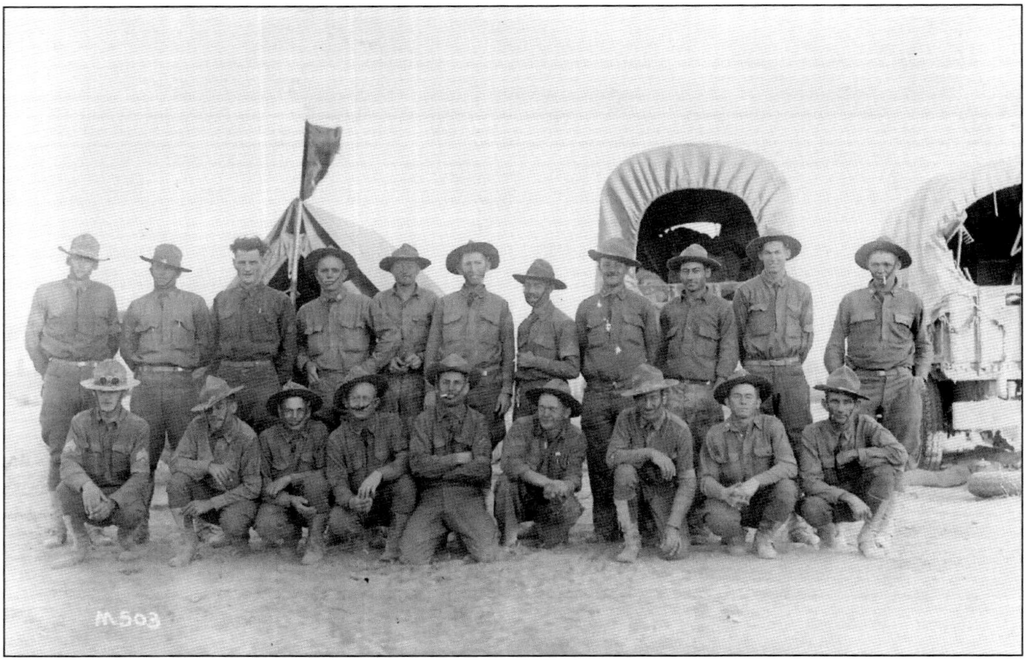

These signal battalion troopers are training in a field near Marfa, Texas, in 1927. Aside from their horse-drawn wagons, their wireless equipment had begun to improve. They could also generate sufficient electrical power to run the radios and even provide lighting. But as with today, field duty was rough and primitive.

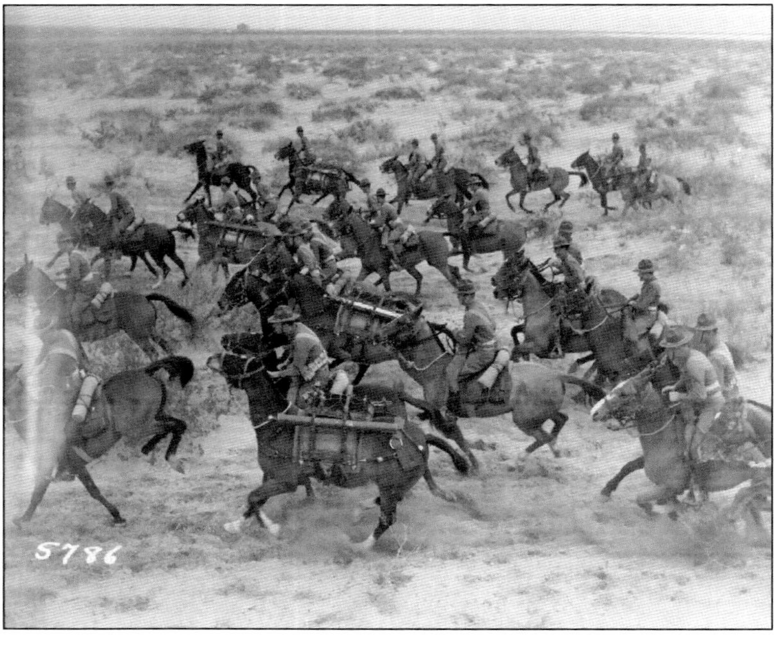

All the cavalry's equipment had to be carried on a horse or mule. This is a cavalry mortar platoon in action, with mortars, bipods, baseplates, and sights tied onto horses. This allowed for mortars to be employed in very rough country, without having to manpack them into place.

Most units had dogs as mascots. This is Sergeant Smokey, the mascot of the 1st Signal Troop of the 1st Cavalry Division, in 1929. He had the run of Fort Bliss, until someone off post shot him. He managed to make it back to his barracks before he died.

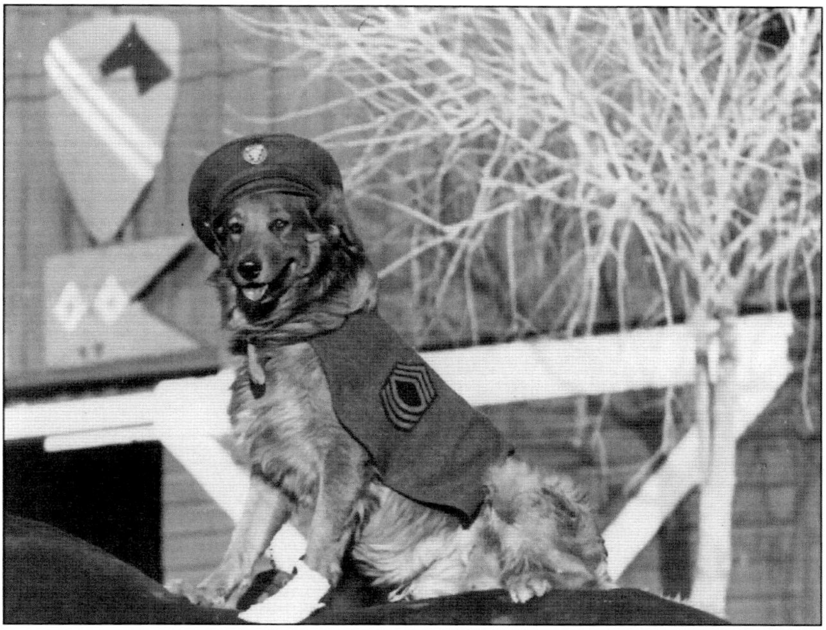

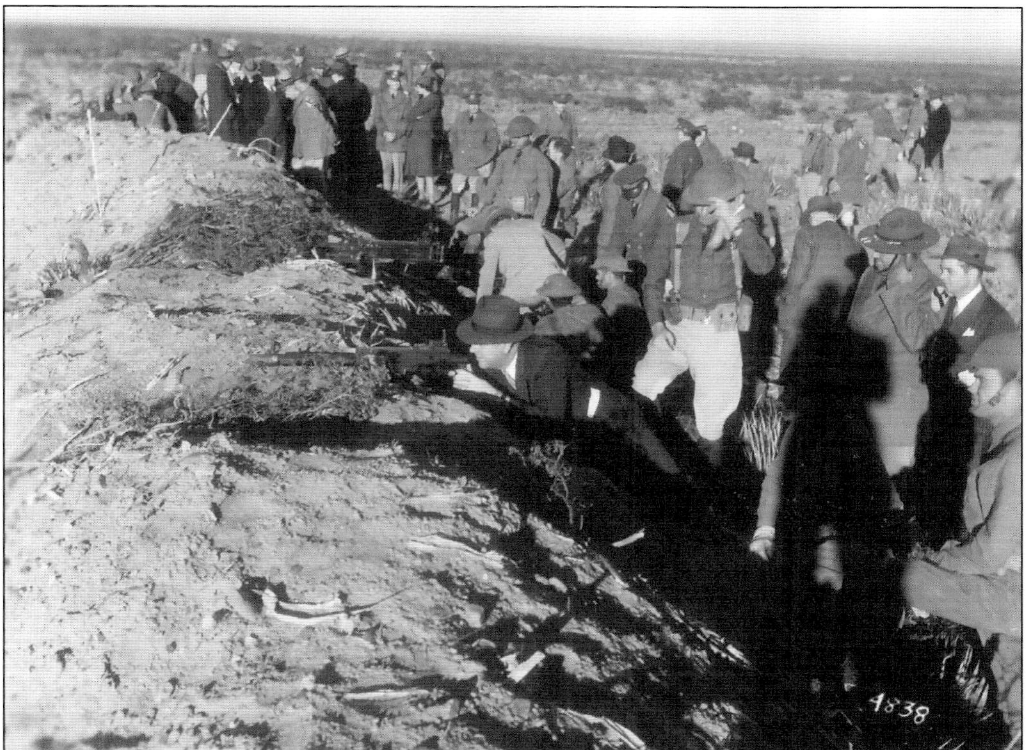

Cavalry officers demonstrate the M1919 light machine guns to local El Paso dignitaries on Dona Ana Range in 1939. Some officers wear the campaign hat and some the M1917 steel "soup plate" helmet. Fort Bliss maintained a very close relationship with the townspeople, as they do today.

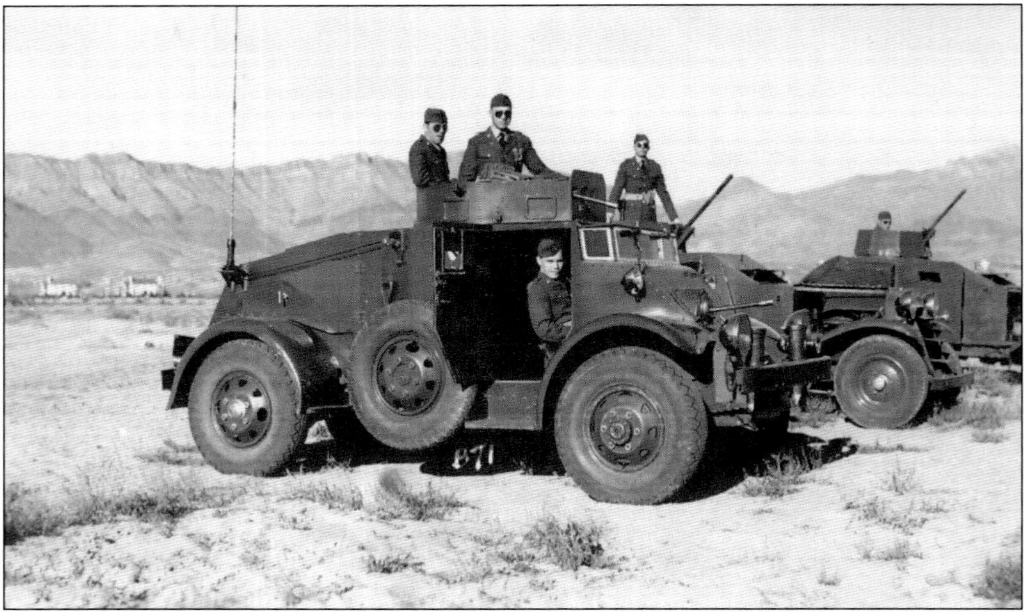

With the continuation of mechanization in the 1930s, a number of experimental armored cars came to Fort Bliss for testing. This is the Marmon Herrington T11 armored car in 1938. One of several experimental armored vehicles, this one had a rear-mounted engine and a turret with a machine gun. This model was later discarded in favor of other models. There are two other experimental vehicles behind this one.

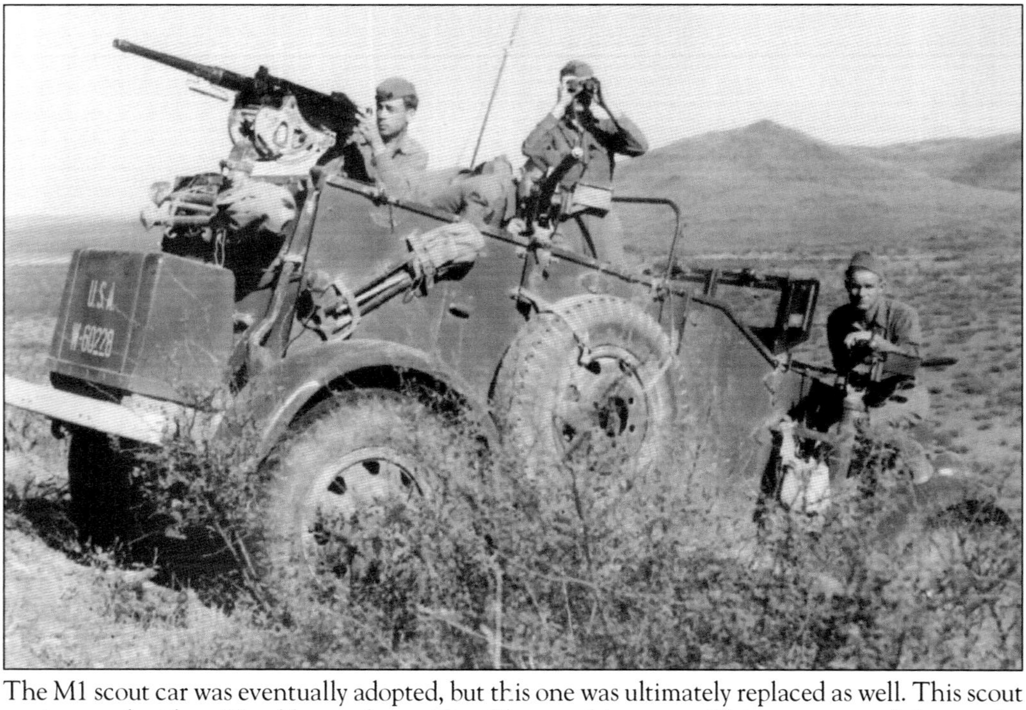

The M1 scout car was eventually adopted, but this one was ultimately replaced as well. This scout car is armed with a .50-caliber and two .30-caliber machine guns. This was a part of the cavalry division test near Toyahvale, Texas, in 1939. These cars were judged deficient because they could not travel where the horse could go.

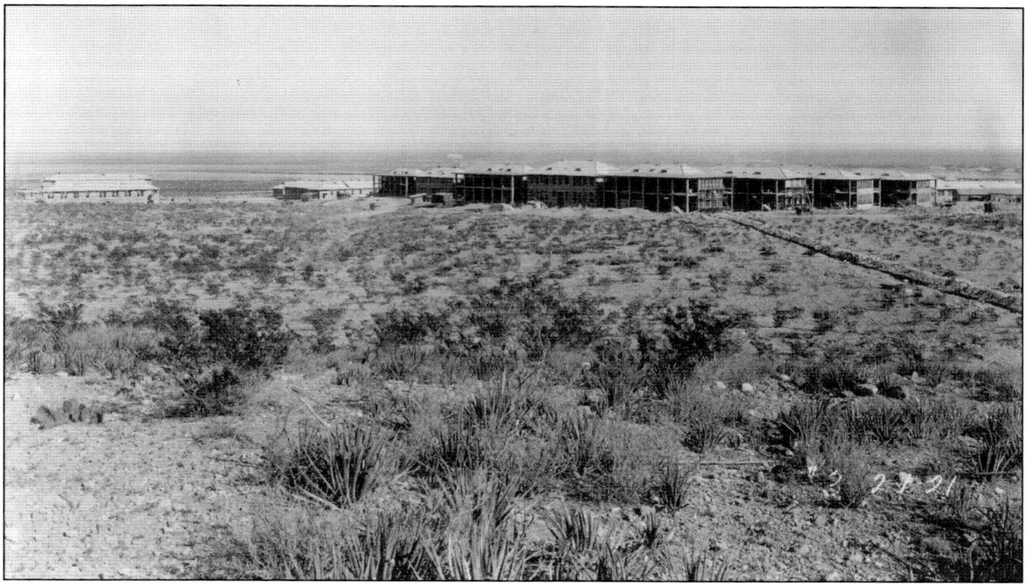

By the time the 1st Cavalry Division was activated, the post hospital had become too dilapidated and crowded. The Army surgeon general acted to fund and build a new and larger hospital. This would be able to cover all the southwestern Army posts. In 1920, construction began on a new facility with a maximum capacity of 414 beds.

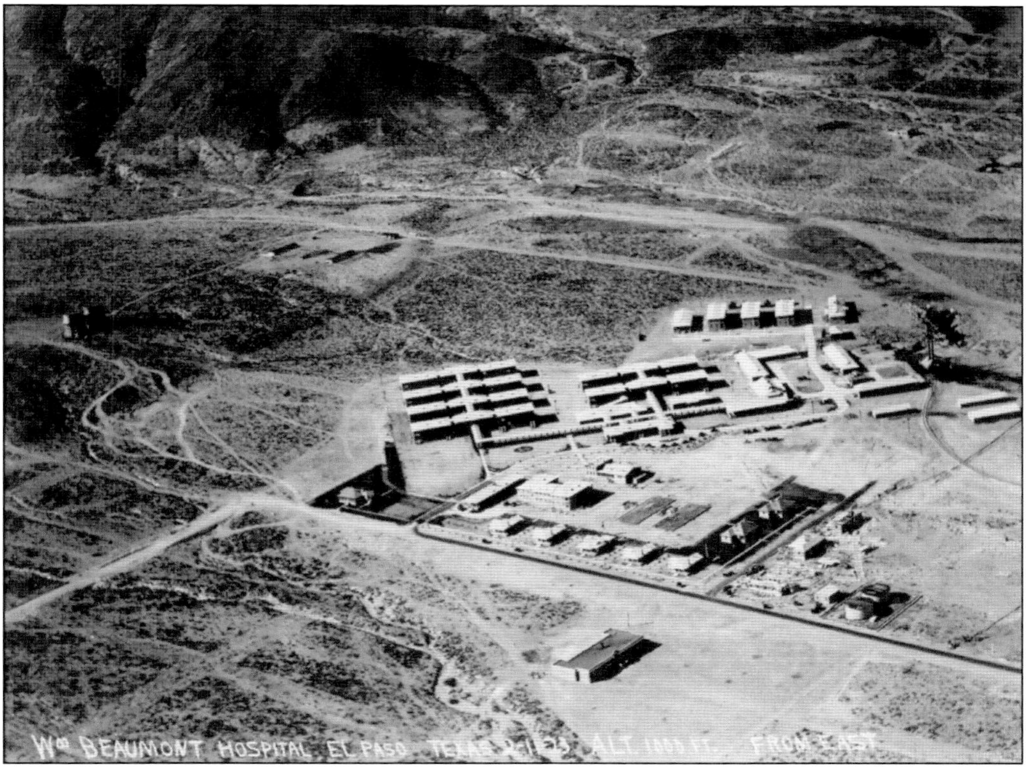

In July 1921, William Beaumont General Hospital opened. It was not complete for another year, but the new facility could handle many more patients and had all the newly developed medical technology. It even had a chicken farm to provide meals for patients and staff.

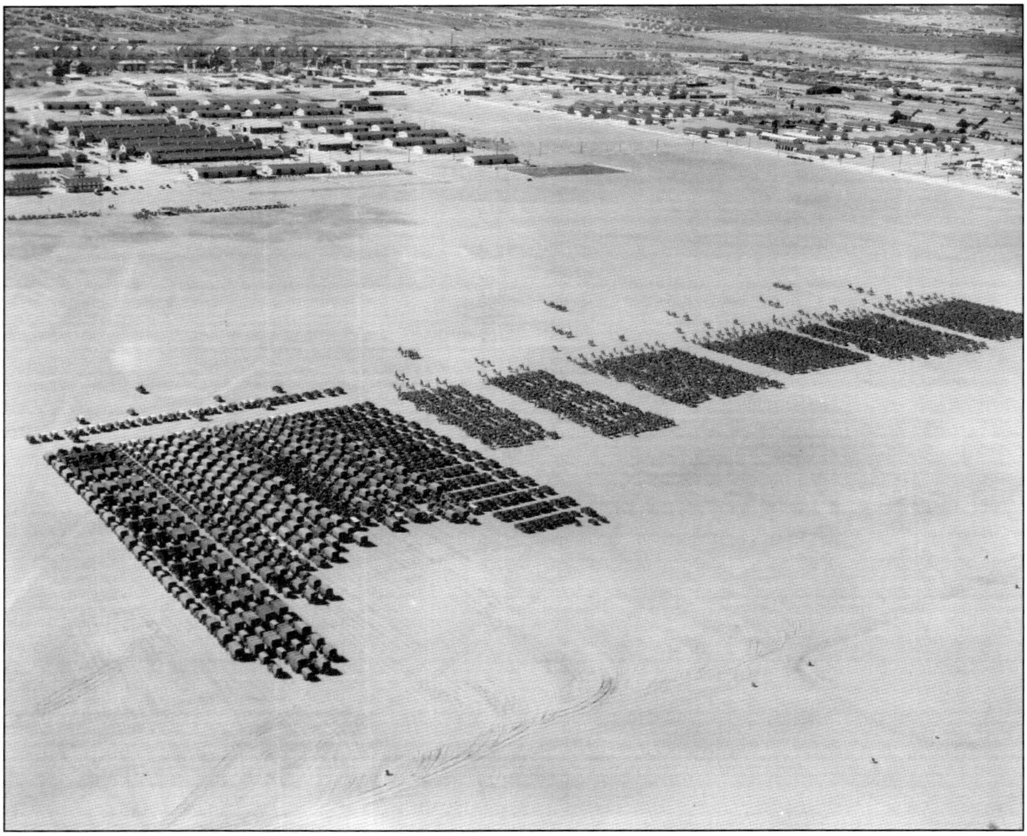

With the 1st Cavalry Division fully established on Fort Bliss, by 1939 it was a very large formation. Although it was a horse cavalry division, resupply in the field could be done by motor vehicle. Each unit had a complement of trucks to carry supplies and fodder for the horses. In 1937, the division had more than 16,000 soldiers.

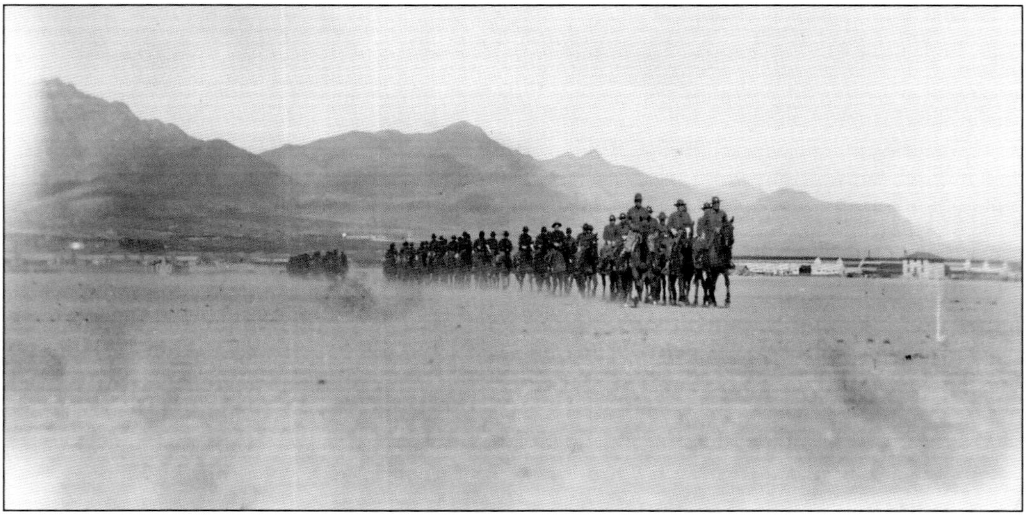

The 7th Cavalry Regiment parades east of the post, with the Franklin Mountains in the distance. The time of horse cavalry was approaching its end. Mechanization was moving fast, and potential adversaries in Europe had adopted mobile warfare concepts to the exclusion of the horse.

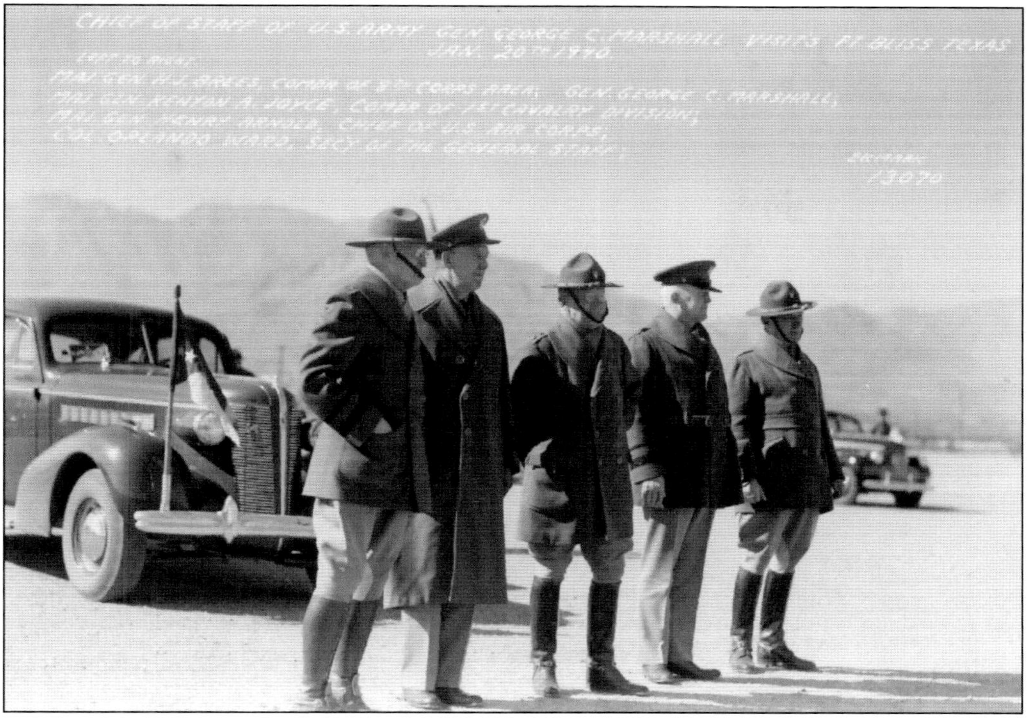

As it became likely that the United States would enter World War II, the Army's senior leadership started visiting Fort Bliss to inspect the cavalry training. Here, Maj. Gen. George C. Marshall, Army chief of staff, reviews division training east of the post. With him are Maj. Gen. Kenyon Joyce, commander of the division; Major General Arnold, Air Corps chief; and Col. Orlando Ward, chief of the general staff.

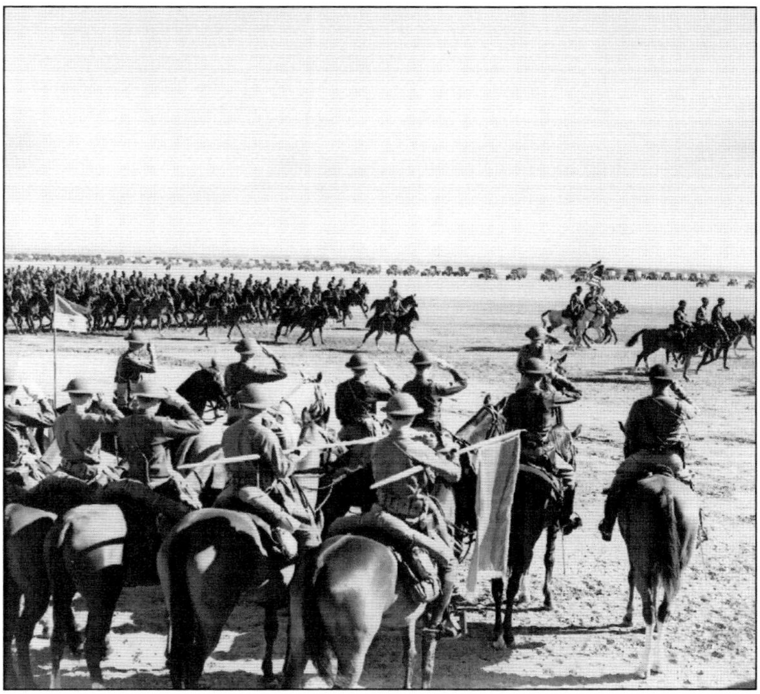

The mature cavalry division passes in review at a gallop in 1939. The division staff and the cavalry troopers are wearing the M1917 steel helmet from World War I. They have finally removed the old-style campaign hat but still retain it for noncombat wear. Very quickly, this helmet would be turned in for the M1 steel helmet with liner, which would be worn for over 40 years.

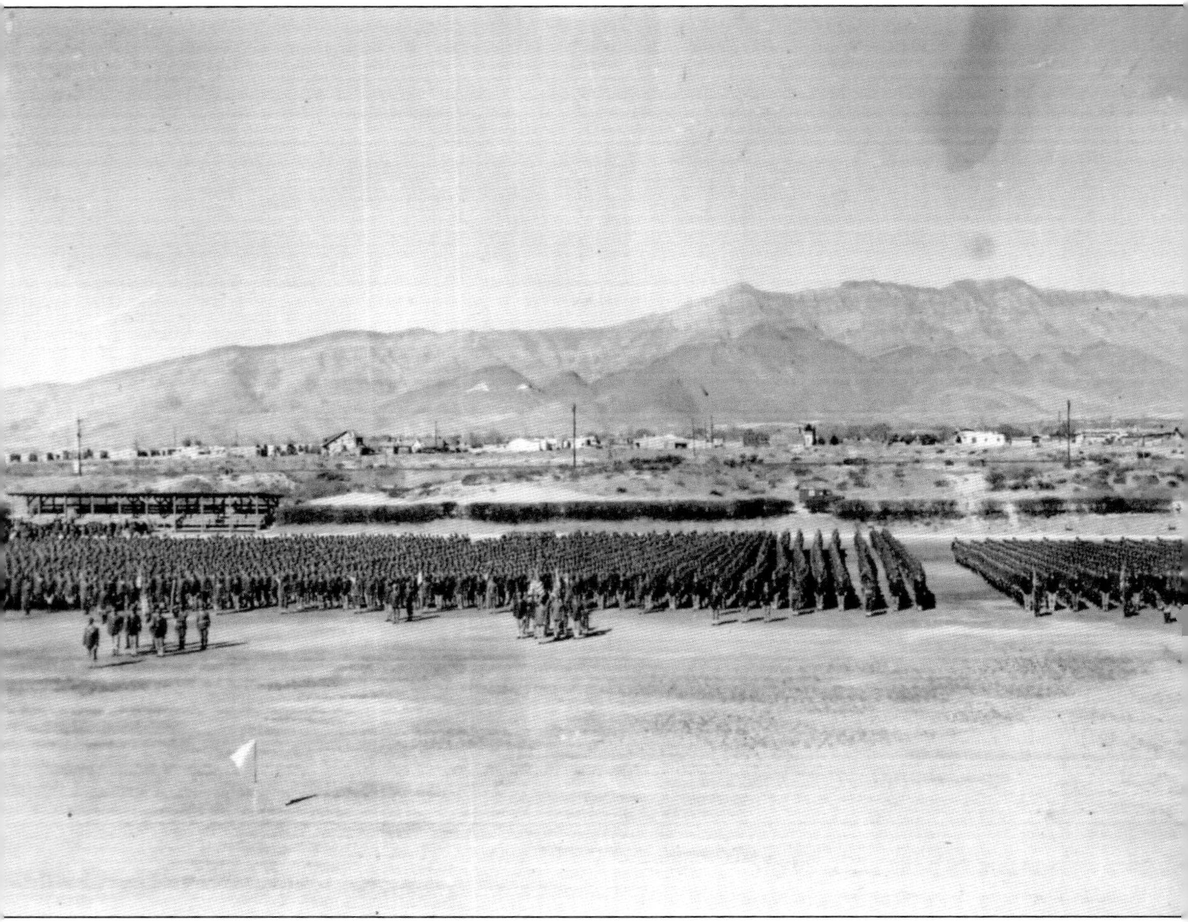

In 1943, the 1st Cavalry Division formed on Armstrong Field to formally dismount. The horses and mules were all sold to local farmers and ranchers. The division was reorganized as an infantry division and was ordered to the southwest Pacific theater to serve under Gen. Douglas MacArthur. This ended the horse cavalry period at Fort Bliss. (Courtesy 1st Cavalry Division Museum.)

Five

Airspace and Rockets
Air Defenders and the Cold War

World War II brought the US Army Antiaircraft Training Command to Fort Bliss. By 1944, all antiaircraft training was consolidated on the post. When World War II ended, the antiaircraft training command at Fort Bliss almost shut down completely. But Operation Paperclip, the German V2 rocket project, came to Fort Bliss in 1945. This injected new life into the post, with its clear weather, open ranges, and proximity to the bombing ranges on White Sands Proving Ground. The Army revitalized the antiaircraft artillery units at Fort Bliss, forming the 1st Antiaircraft Guided Missile Battalion. This later expanded into a regiment and then a brigade as the rocket testing proceeded.

The advent of the Cold War brought about more attention to the air defense mission. The United States Army responded to the Soviet Union's military developments by forming the Army Air Defense Command with its defense installations near major cities and military installations. The antiaircraft guns were quickly supplanted with the new Nike Ajax missile in 1952, followed by the Nike Hercules in 1958. Fort Bliss became the training post for the soldiers manning these systems, and the Army Air Defense Center was born.

The post expanded dramatically to 1.12 million acres to allow for missile training and testing. Other units came to the post as well, with the 3rd Armored Cavalry Regiment arriving in 1972 after leaving the Fort Bliss area in 1861. The post operated Army basic training for new soldiers several times to provide trained soldiers in different conflicts. The Army established a new hospital, improved soldier barracks and family quarters, and maintained its close relationship with El Paso. For a while, the US Air Force maintained a Strategic Air Command base, turning it back over to the Army in 1962.

After 69 years of air defense artillery at Fort Bliss, the Department of Defense implemented Base Realignment and Closure legislation in 2005. The Air Defense Center moved to Fort Sill, Oklahoma, and the 1st Armored Division arrived from Germany. But the old post endures, dating back to 1849.

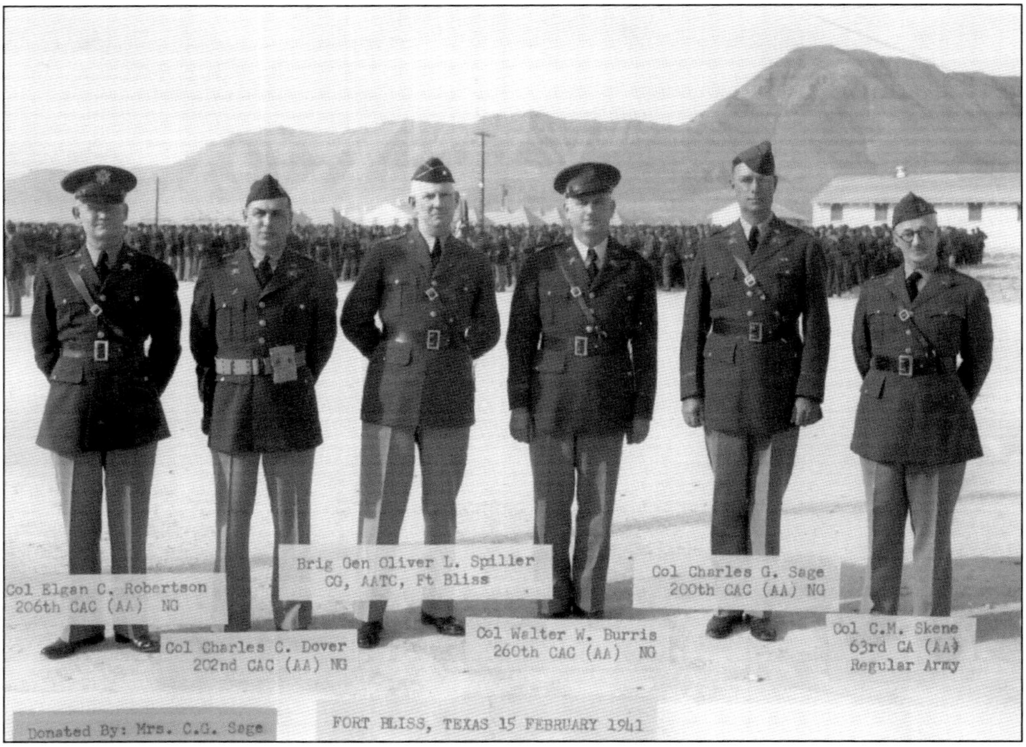

As the United States' entry into World War II approached, the Army started mobilizing and training forces. With a substantial aerial bombardment threat, antiaircraft training bases were positioned at seven posts. Accordingly, five coast artillery antiaircraft regiments came to Fort Bliss between 1940 and 1941. These troops were housed in tent camps on Logan Heights, northwest of the post. (Picture donated to the Fort Bliss Museum by Dorothy Haynes Sage.)

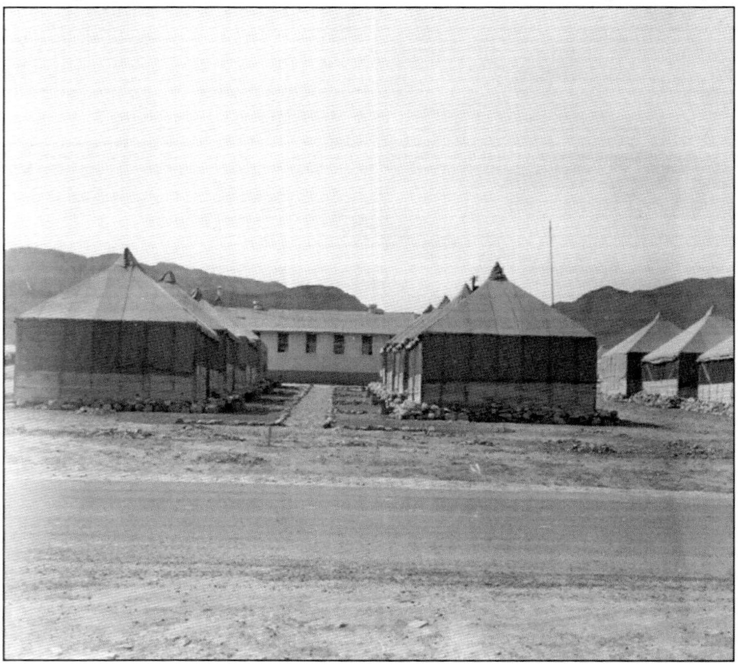

The arrival of so many troops to Fort Bliss so quickly put a strain on housing. Tents with wooden siding and floors were the first quarters. These were heated with butane stoves. All buildings and tents had electric lighting. All mess halls, kitchens, latrines, and bathhouses had hot and cold running water. Prefabricated hutments soon replaced the tents. The 1,000-acre Logan Heights area filled up fast.

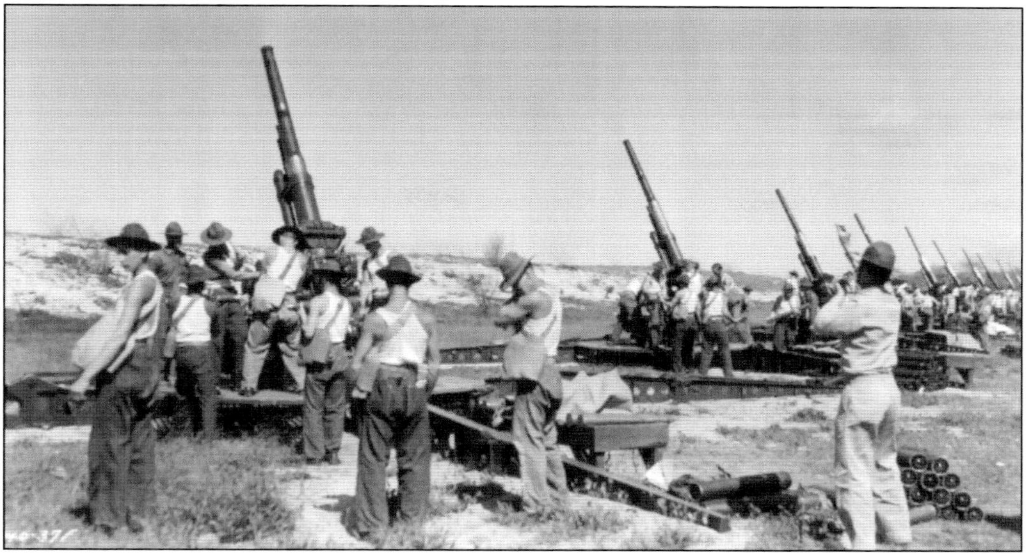

Antiaircraft training began in earnest in 1940. Fort Bliss was an ideal training location, with clear skies and open range areas. At least 10 three-inch guns were on line at Oro Grande Range in 1940. Troops are wearing campaign hats and are shirtless in the heat. Troops also carry chemical protective masks. The gun at left has just fired.

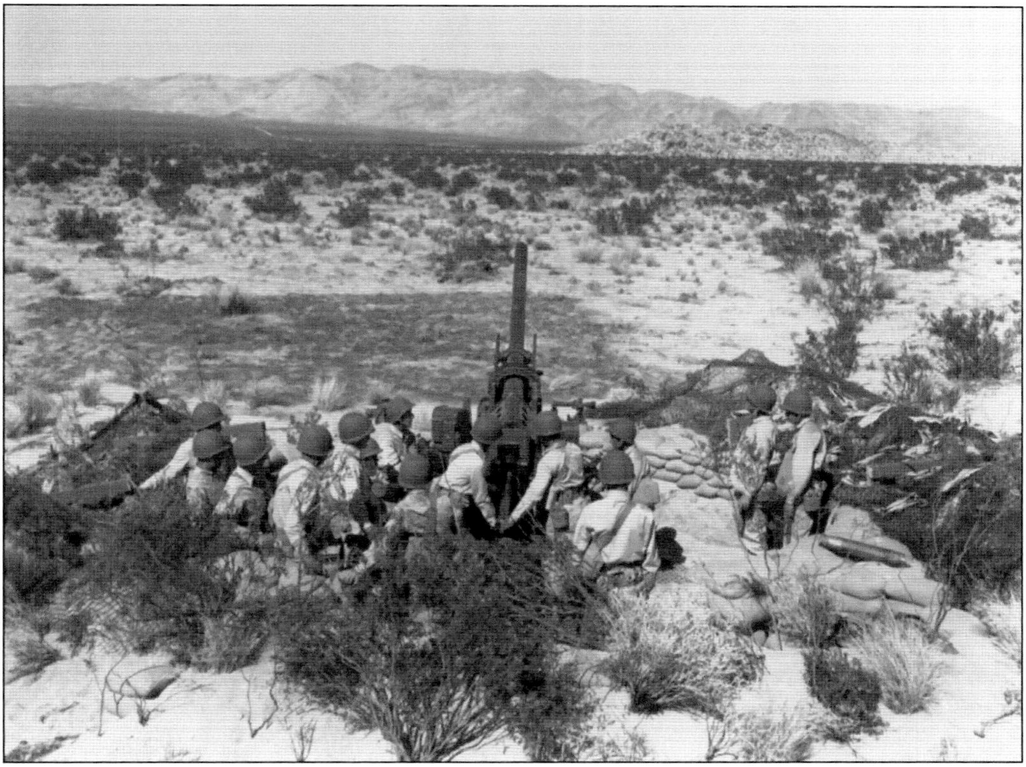

A 90-mm gun crew conducts live fire training on Oro Grande Range in the 1940s. Tow planes from the 10th Observation Squadron at Biggs Army Airfield towed aerial targets for crews to shoot at. By 1943, antiaircraft strength in the Army had increased 1,750 percent. Fort Bliss would eventually be the primary training post for all antiaircraft troops in World War II.

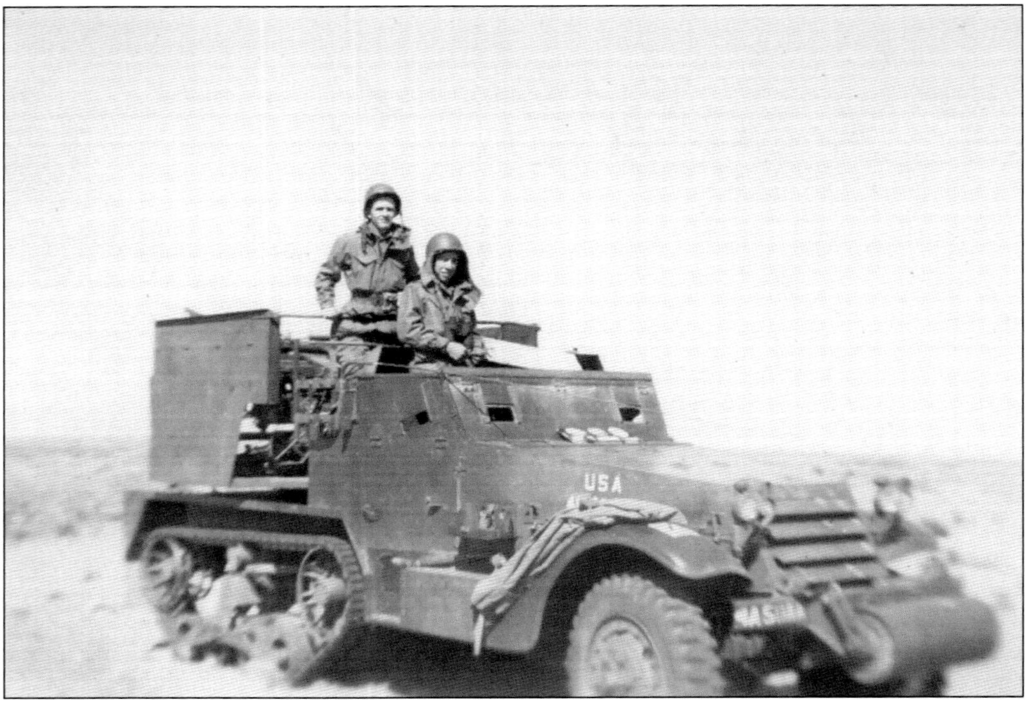

Along with heavy antiaircraft guns, Fort Bliss troops trained self-propelled short-range antiaircraft units. This is an M15 antiaircraft half-track. It mounted a 37-mm antiaircraft gun with a .50-caliber machine gun on either side. The half-track could move on roadways at a top speed of 45 miles per hour. The desert conditions at Fort Bliss proved to be an ideal training area for troops heading to combat in North Africa.

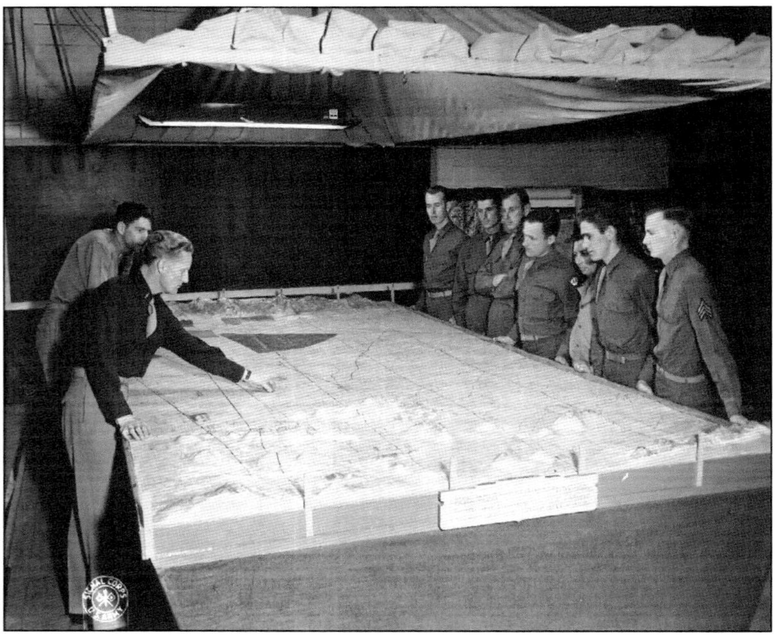

Not all training was outdoors. Some troops trained indoors using terrain models. Two lieutenants train sergeants on map and terrain features using a relief model. This could be a time-saver, and it was certainly easier than trying to move around the training area or flying above it.

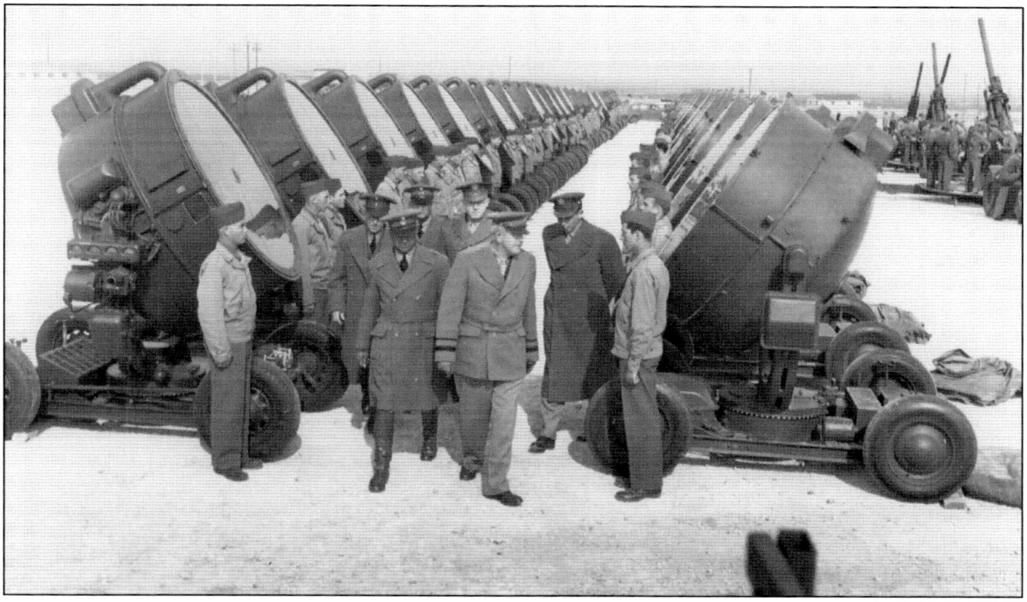

Unit inspections were the norm during this period. Here, the Antiaircraft Training Center commander inspects a searchlight battery with a Mexican army officer along for the experience. Mexican officers were welcome at Fort Bliss during the 1930s and 1940s for inspection tours and joint training. These 60-inch searchlights produced 800 million candlepower and could cast a beam up to 35 miles.

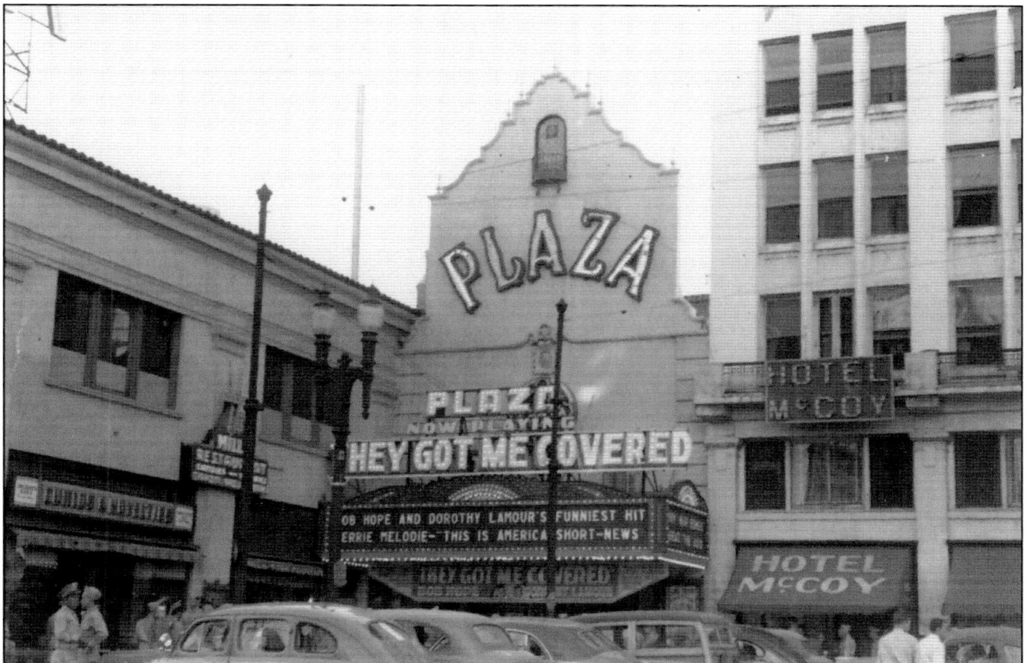

Troops had many diversions, some wholesome, some not. Soldiers who were not on duty could go to the movies in the Plaza Theater in downtown El Paso. But to leave the post, they had to be in uniform with a pass signed by their commander or first sergeant. Military police patrols would detain soldiers and return them to the post if they could not show a valid pass.

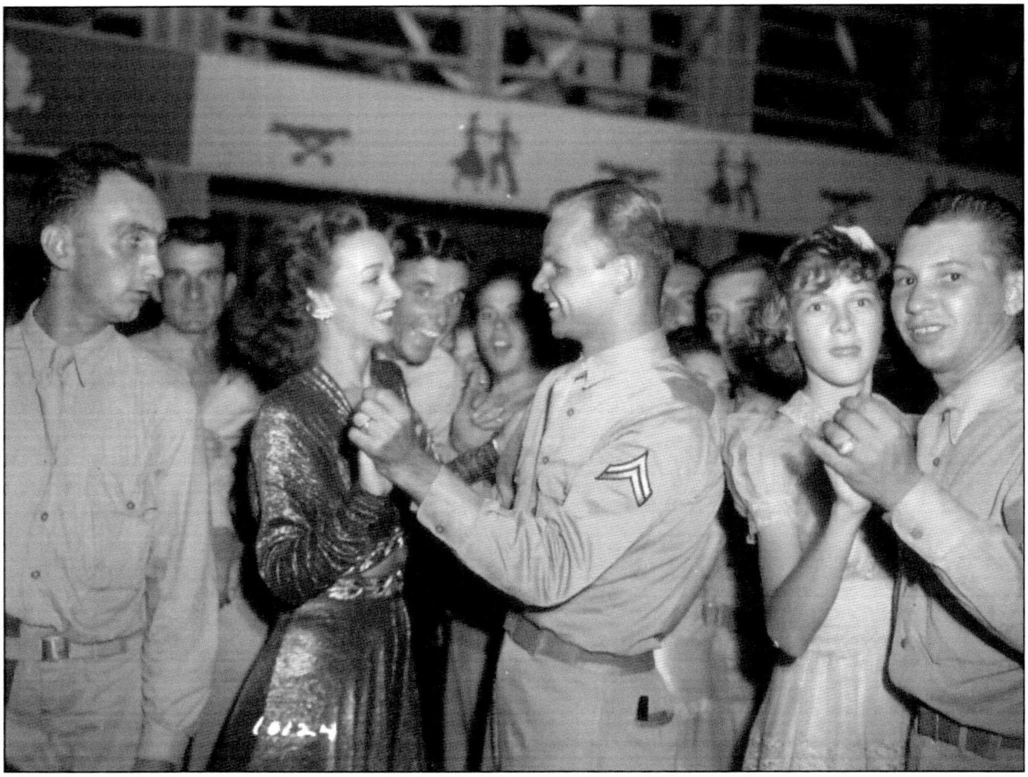

Service clubs provided diversions for off-duty troops. The Army Special Services organization regularly sponsored dances, inviting young women from off post to attend. The club would also host traveling United Service Organizations (USO) shows that included big-name movie stars and celebrities. Here, soldiers vie to dance with movie star Carol Landis in the Logan Heights club.

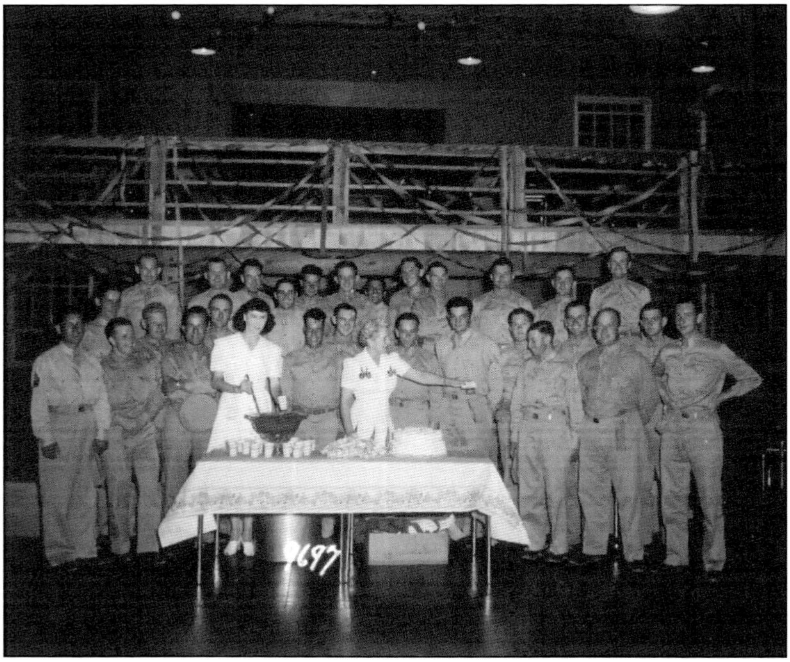

Soldiers in khaki are treated to a birthday party during a service club dance in 1942. Hostesses came from off post to serve as partners and companions during a dance. Many women married antiaircraft soldiers who trained at Fort Bliss before and during the war.

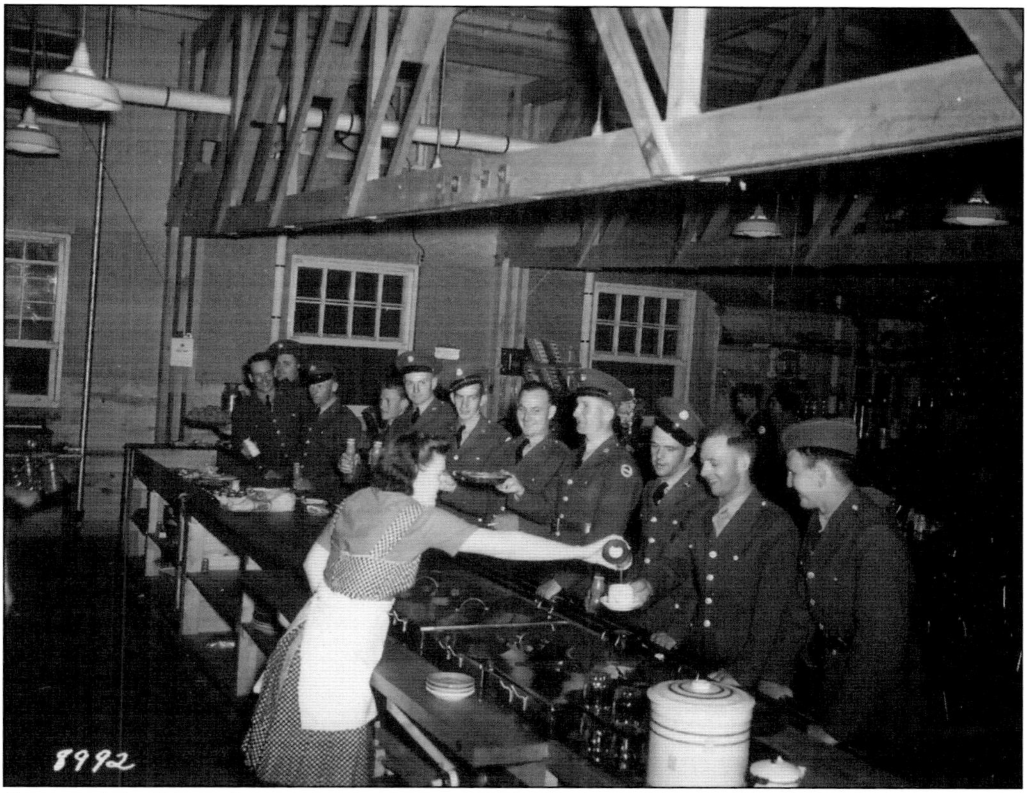

Service clubs featured many wholesome diversions besides dances or hops. A soldier could get a good meal at a very reasonable price, play cards in the cardroom, read books and magazines in the club library, or just relax and listen to the radio. With a very busy training schedule and much field duty, the service club was a refuge from the tent or hutment.

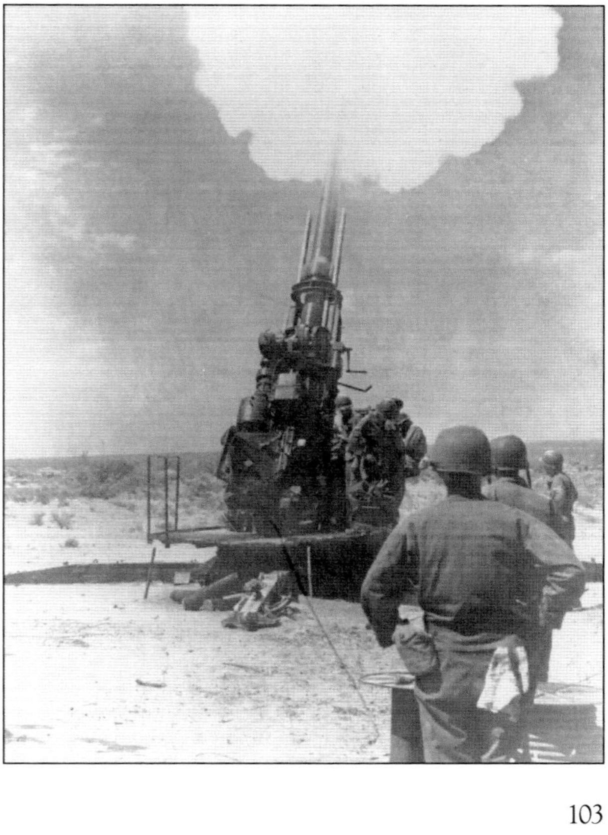

A 120-mm antiaircraft gun known as the stratosphere gun is fired at Oro Grande Range. Developed before World War II, the stratosphere gun was the Army's heaviest antiaircraft gun with a maximum firing range of 81,000 feet. After World War II, this gun was deployed to protect industrial centers, big cities, and other potential air targets during the Cold War.

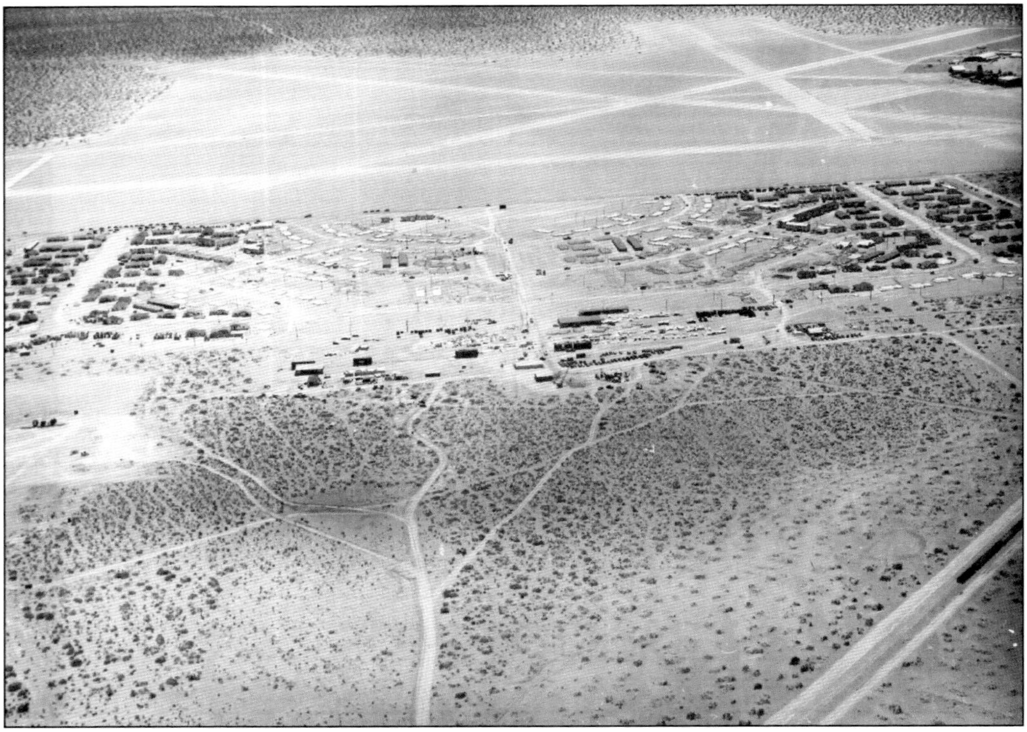

Biggs Army Airfield in 1942 was an important transit point for Army Air Force aircraft. This was the home of observation squadrons, antiaircraft target-towing aircraft, and even bombers. Biggs Army Airfield appears very primitive in this photograph but improved tremendously as the war went on.

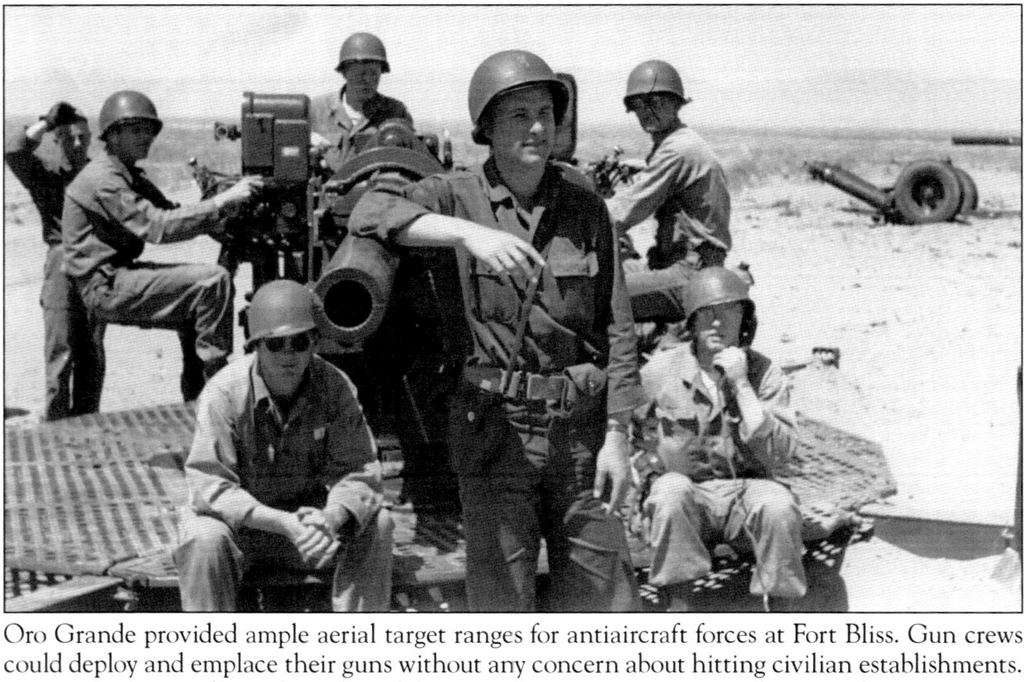

Oro Grande provided ample aerial target ranges for antiaircraft forces at Fort Bliss. Gun crews could deploy and emplace their guns without any concern about hitting civilian establishments. Here, a gun crew adjusts the sights of their 90-mm gun in preparation for range firing.

The workhorse radar for World War II antiaircraft crews was the SCR 584 radar. This was a common sight at Fort Bliss during and after World War II. It was the most advanced gun-laying radar of its time, used to aim antiaircraft guns on moving aircraft. This system lasted well into the 1950s.

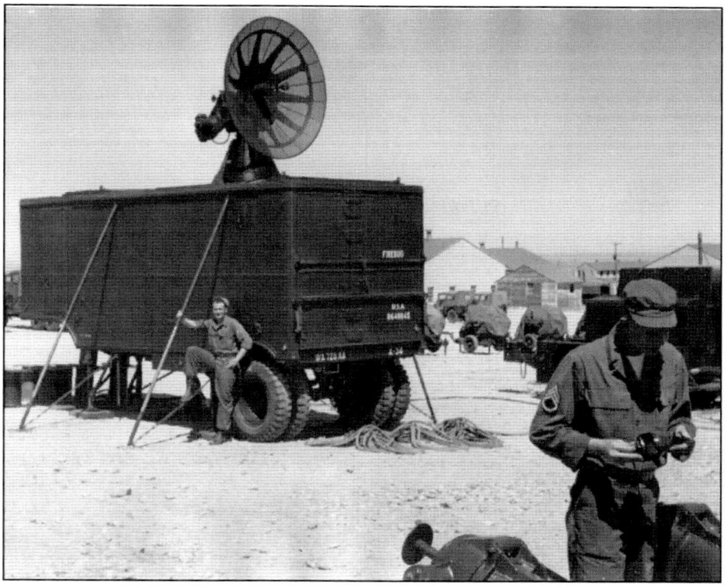

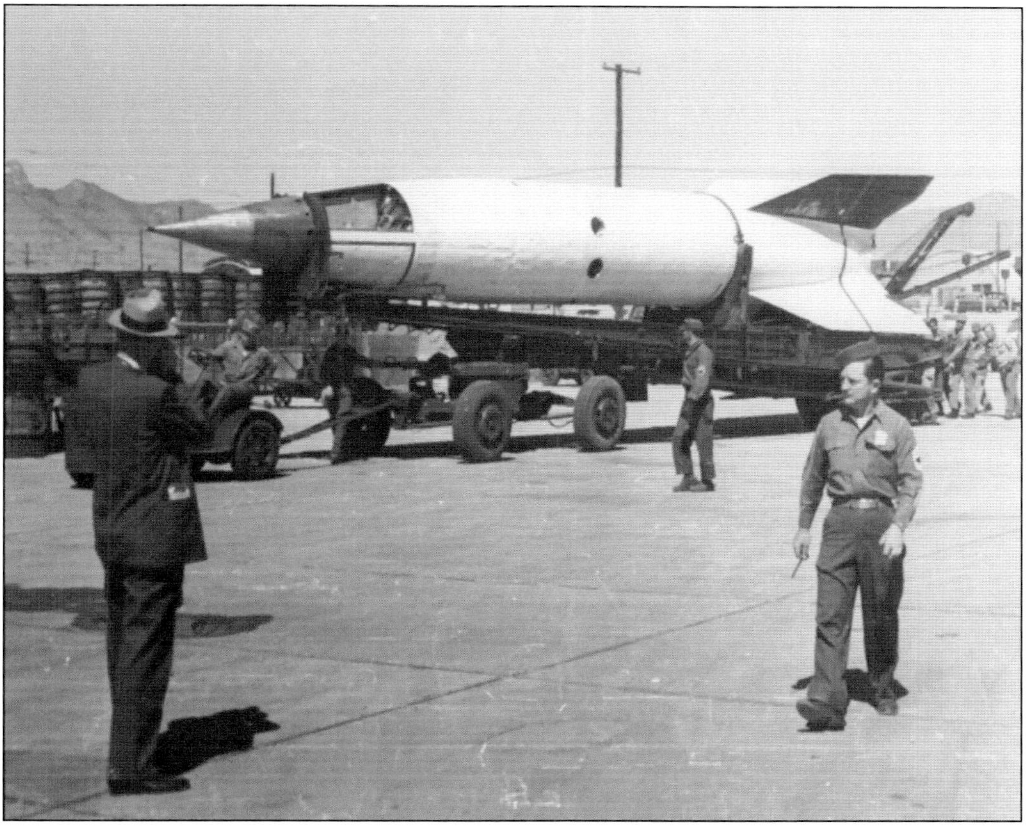

After World War II, the Army Ordnance Department conceived a plan to bring captured German rocket scientists to Fort Bliss to continue developing rockets. In August 1945, three hundred freight carloads of V2 rockets and rocket parts arrived at Fort Bliss. In February 1946, a total of 118 scientists arrived from Germany to continue developing and testing the V2. This was called Operation Paperclip, named for the method of identifying the files of selectees.

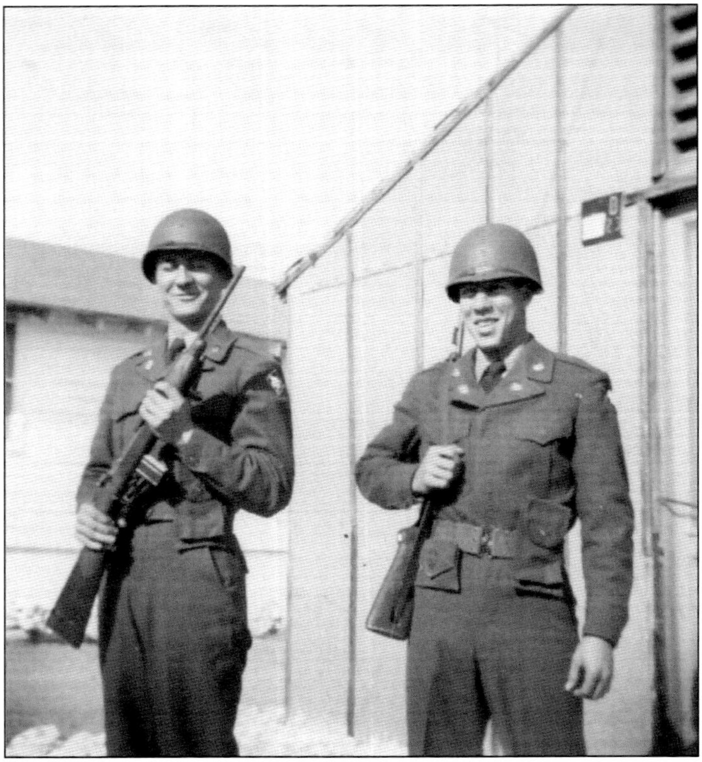

Logan Heights, to the northwest of the main post, was the antiaircraft training area until the 1st Cavalry Division left the post in 1943. The original tent camp established in 1940 was replaced with prefabricated hutments, which were 5–10 degrees cooler than the old tents. These troops prepare for inspection in front of their hutment.

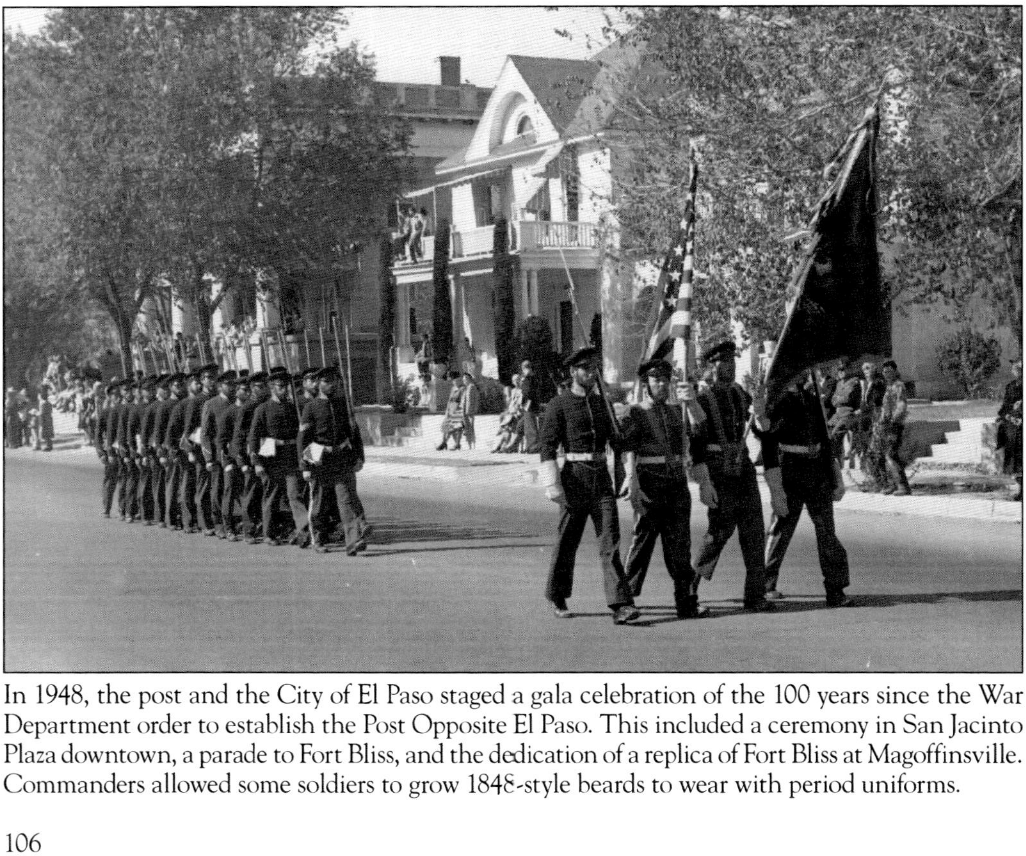

In 1948, the post and the City of El Paso staged a gala celebration of the 100 years since the War Department order to establish the Post Opposite El Paso. This included a ceremony in San Jacinto Plaza downtown, a parade to Fort Bliss, and the dedication of a replica of Fort Bliss at Magoffinsville. Commanders allowed some soldiers to grow 1848-style beards to wear with period uniforms.

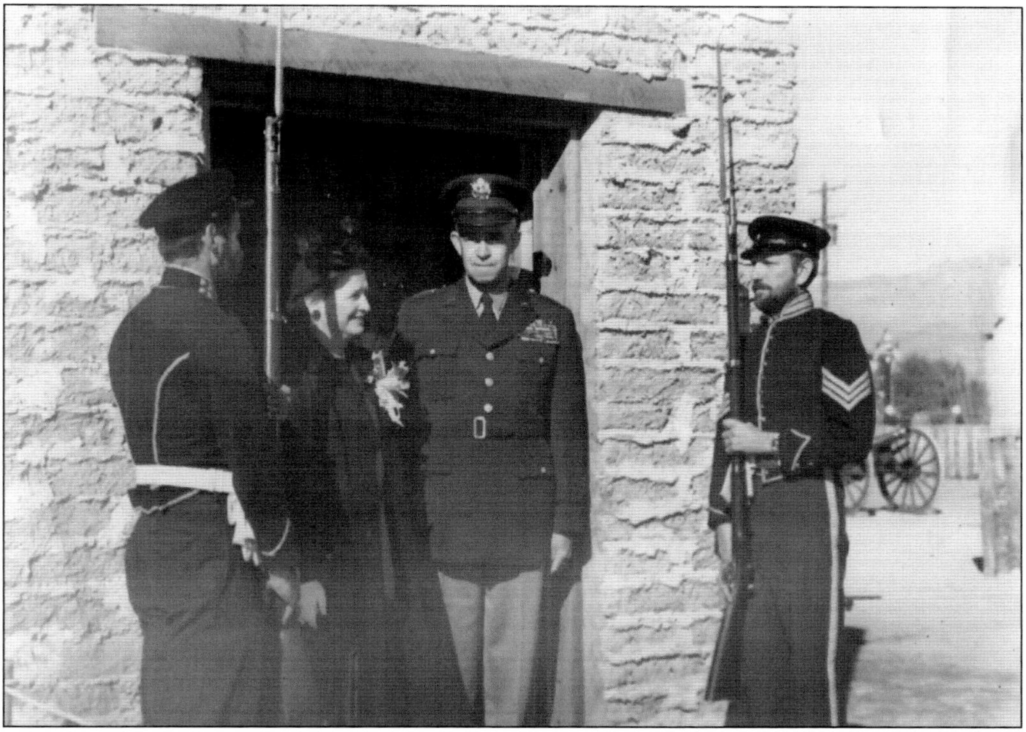

General of the Army Omar N. Bradley and his wife, Mary Quayle Bradley, visit the Magoffinsville replica dedication on Saturday, November 6, 1948. The new replica was made from local adobe and timbers from the Capitan National Forest. It was open to the public November 5–7. It still stands on Fort Bliss and is open to the public.

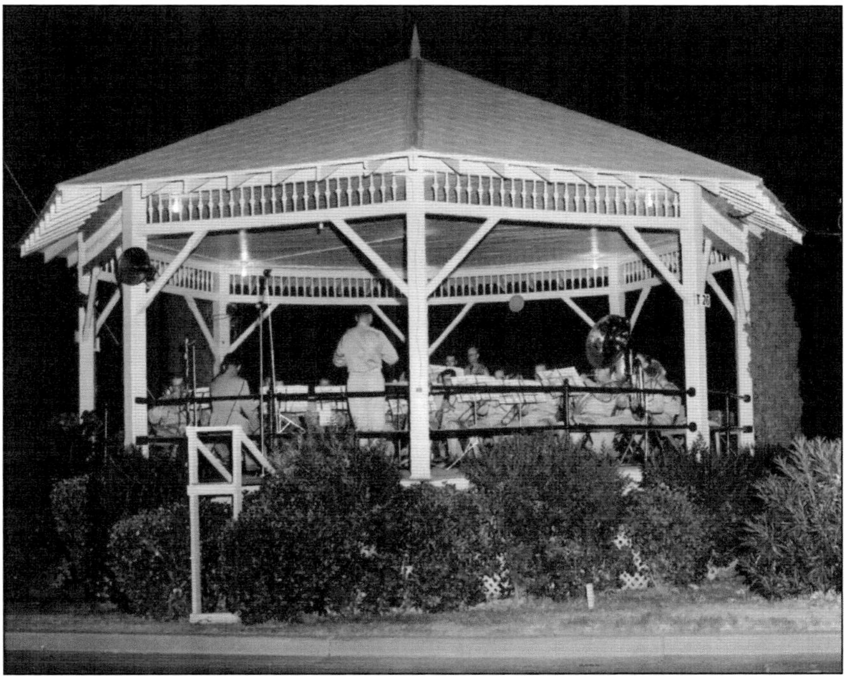

The band gazebo at William Beaumont General Hospital served as an entertainment venue for patients and staff. There was also a gazebo on the main parade field for many years. Here, a band plays in the cool of the evening.

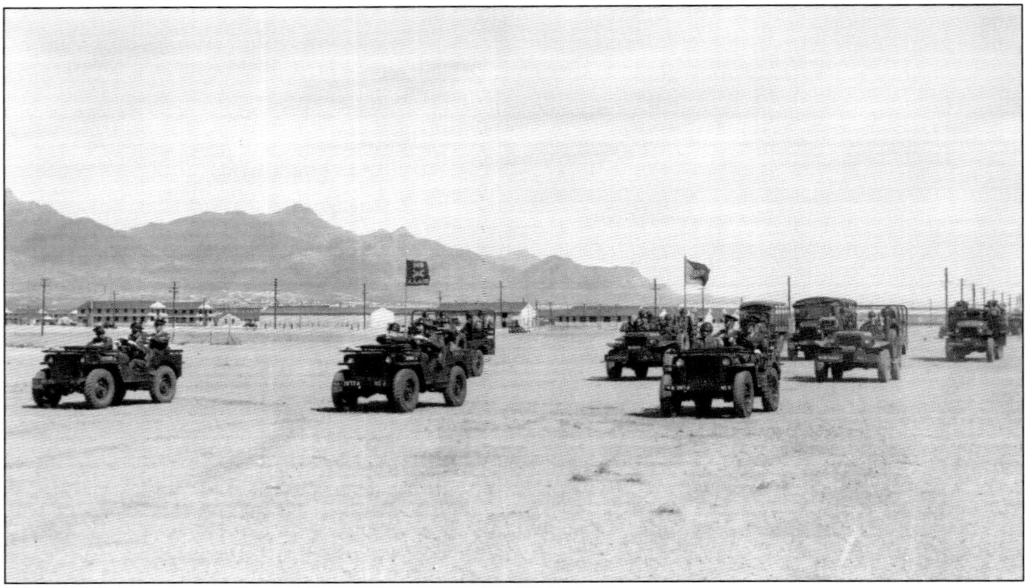

The establishment of Operation Paperclip at Fort Bliss begat the establishment of the 1st Antiaircraft Guided Missile Battalion. This battalion expanded into a regiment and then a brigade as missile development became the post's mission. By 1948, Fort Bliss was the antiaircraft and guided missile center for the Army. Here, units stage a mounted review on April 3, 1948.

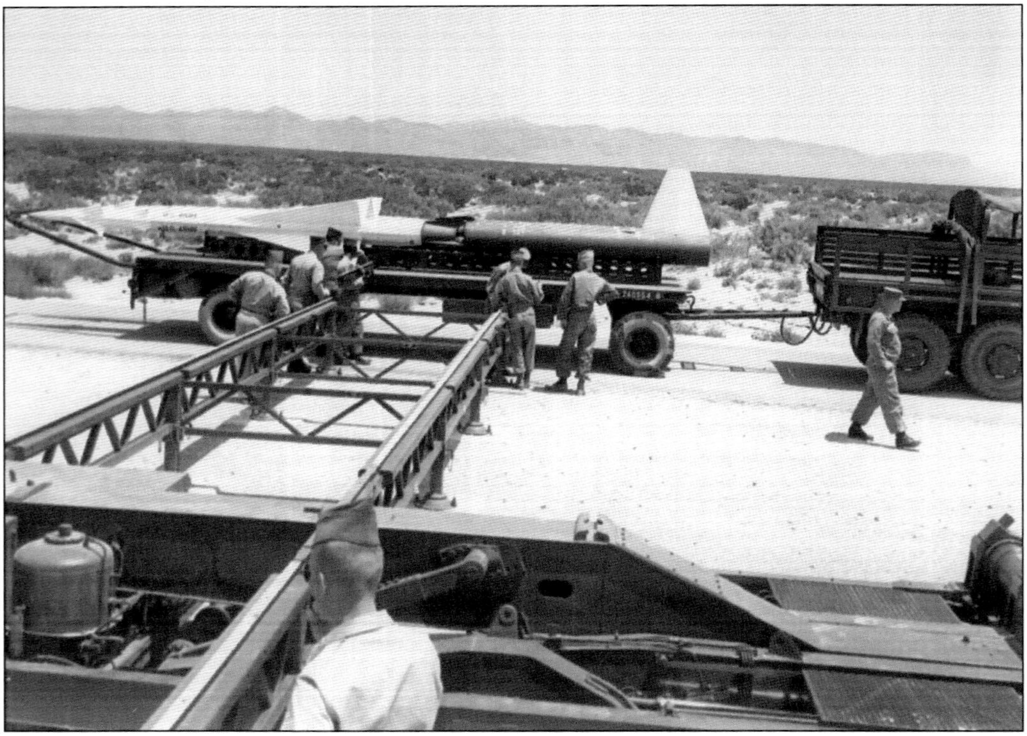

Out of V2 testing came the Army's guided missile program. The Nike Ajax air defense missile was a revolutionary development, making the antiaircraft gun obsolete forever. By April 1952, the Nike Ajax was operationally ready. The Army sent its antiaircraft gun crewmen through a stringent course of training to field Nike Ajax units as part of the Army Air Defense Command.

The Nike Ajax was a two-stage rocket. The booster stage was solid-fueled, boosting the rocket to the appropriate speed before separating. The sustainer stage was liquid-fueled, with JP-4 jet fuel and red fuming nitric acid. The fuel was volatile, necessitating protective clothing for personnel. The heavy rubber suits with head and face protection were no fun to wear at Fort Bliss during the summer.

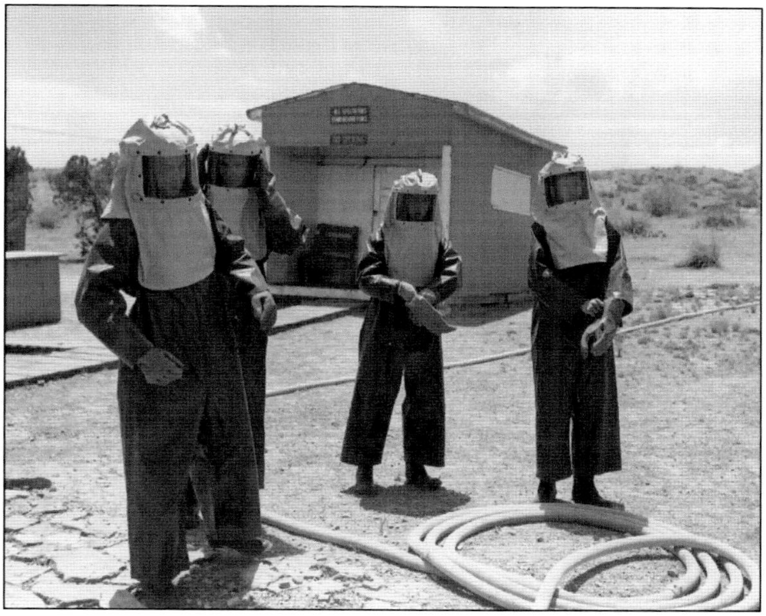

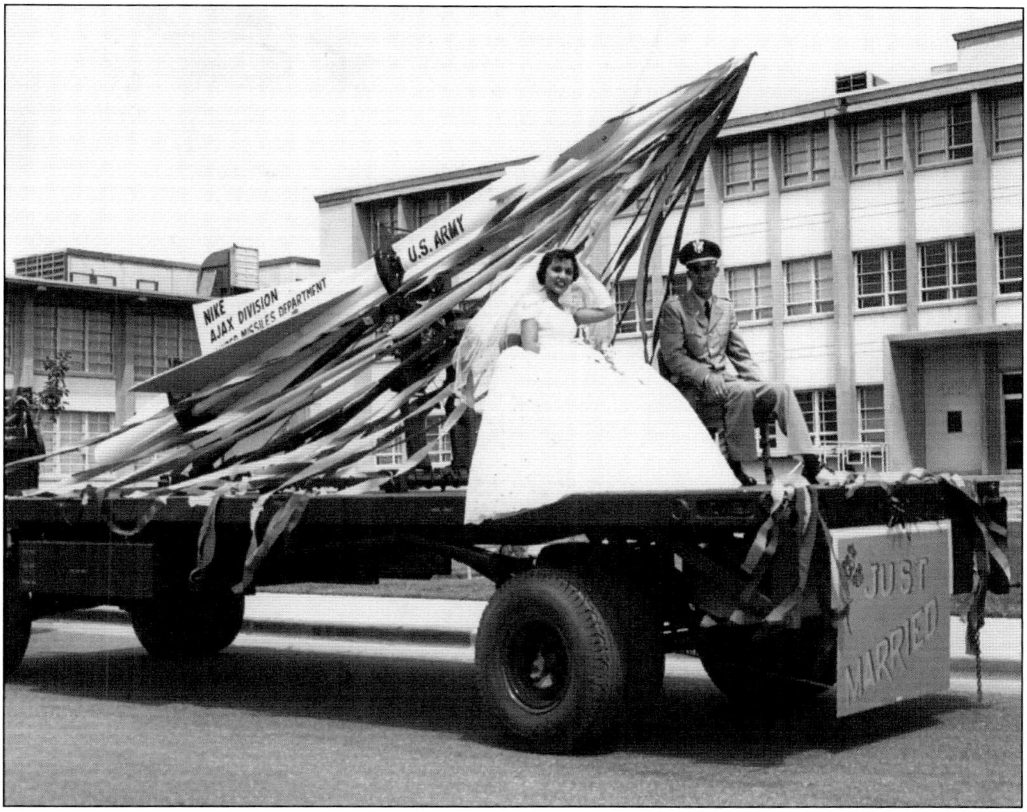

Newlyweds Lt. William Schneider and the former Barbara Carver pose on a Nike missile trailer after their marriage. This was a revival of the old "caisson ride," a ride for newlyweds aboard a rickety gun caisson around the post. Supposedly, after the rough ride the couple's marriage would seem smooth for the rest of their lives.

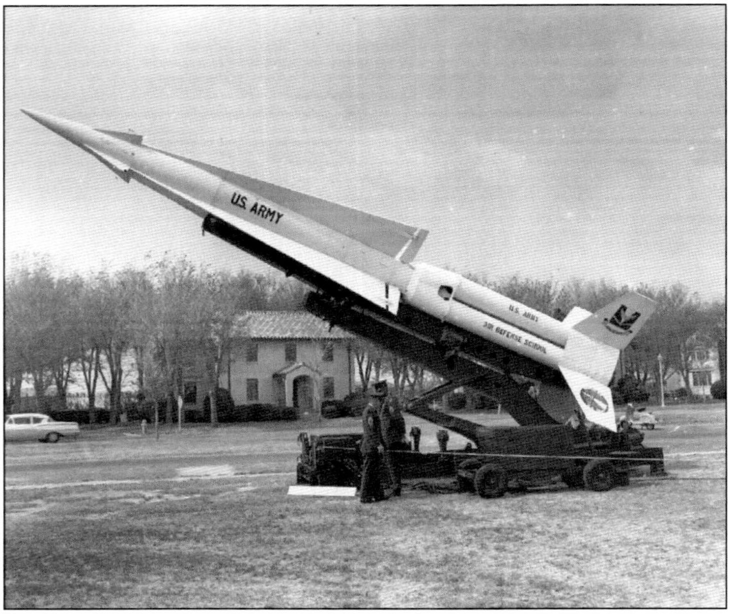

The Army had replaced the Nike Ajax with the Nike Hercules missile by 1958. Where the Ajax was effective against single aircraft at high altitudes, the Nike Hercules could engage and defeat formations of aircraft at double the altitude. It was also completely solid-fueled. Here, a Nike Hercules is displayed on Noel Parade Field. This missile could carry a nuclear or conventional warhead and could also hit ground targets.

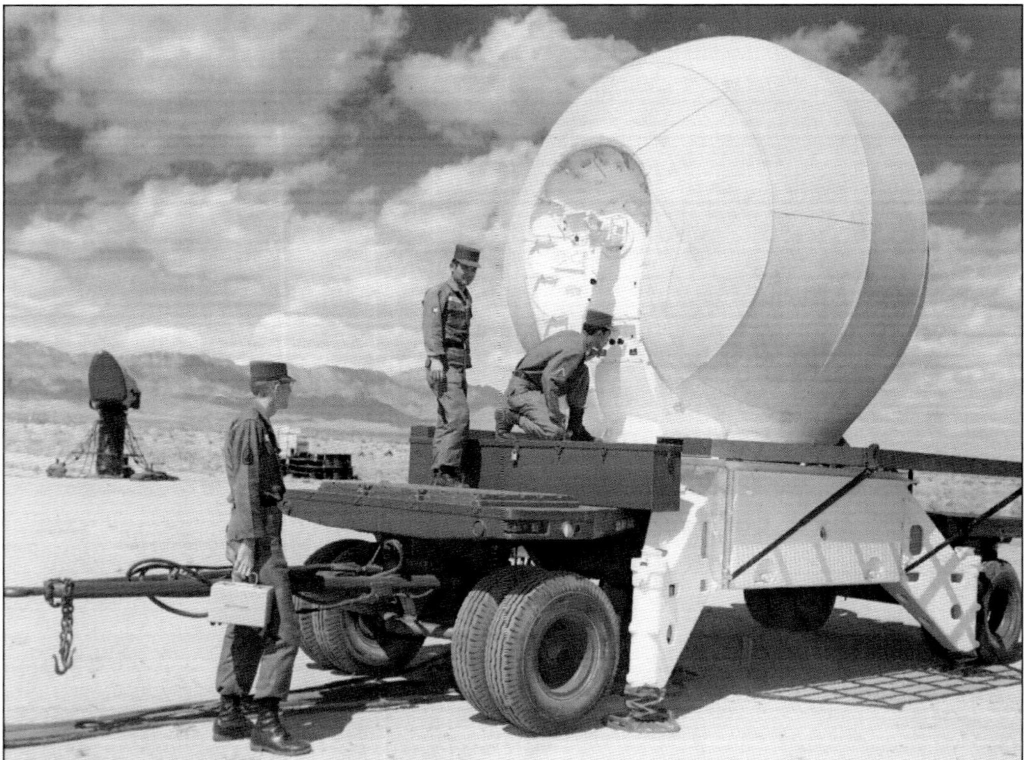

Controlling the flight of the Nike Hercules was the Nike Hercules target tracking radar. These behemoths were mounted on trailers, so they could be mobile. The Nike Hercules system was the backbone of United States air defense until 1965, when Soviet deployment of intercontinental ballistic missiles made them obsolete. Still, the Army retained them in Europe until 1983 when the Patriot missiles replaced them.

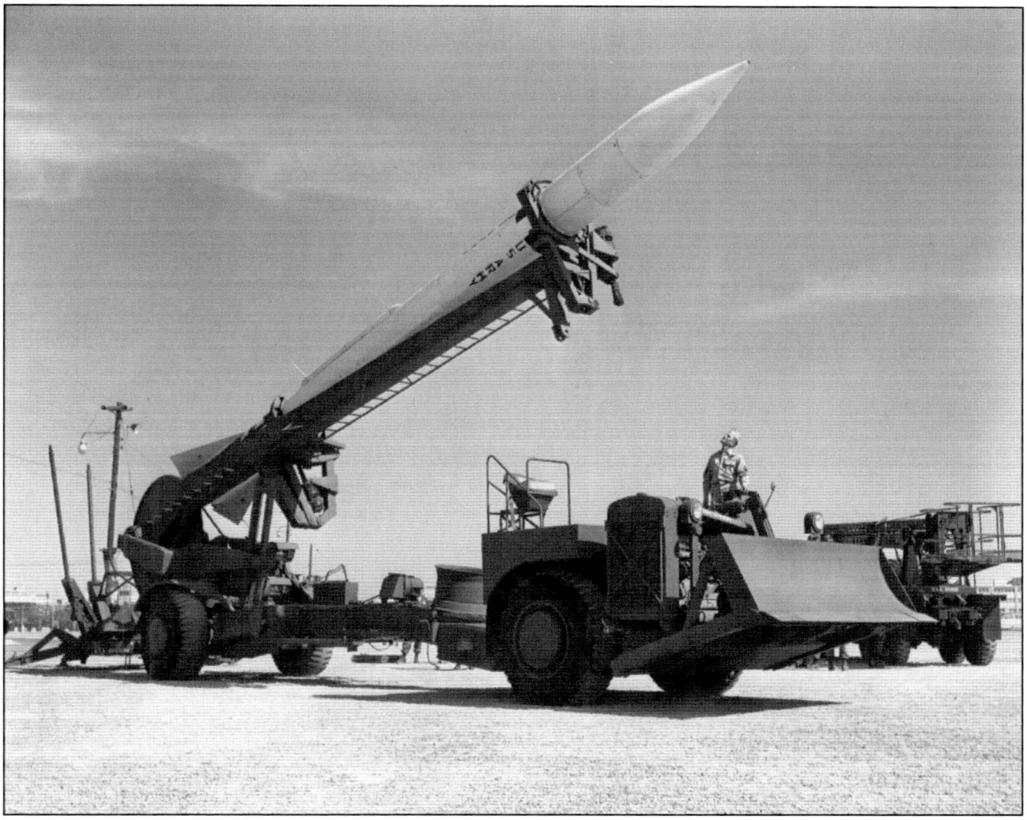

One of the missiles developed from the V2 was the Corporal. This was an artillery ballistic missile that could carry a 1,500-pound warhead over 75 miles. It was a complex system, using liquid fuel and requiring very specialized vehicles like the transporter seen here. It took upwards of two years to train a crewman, after which that crewman could leave the Army for a much better job.

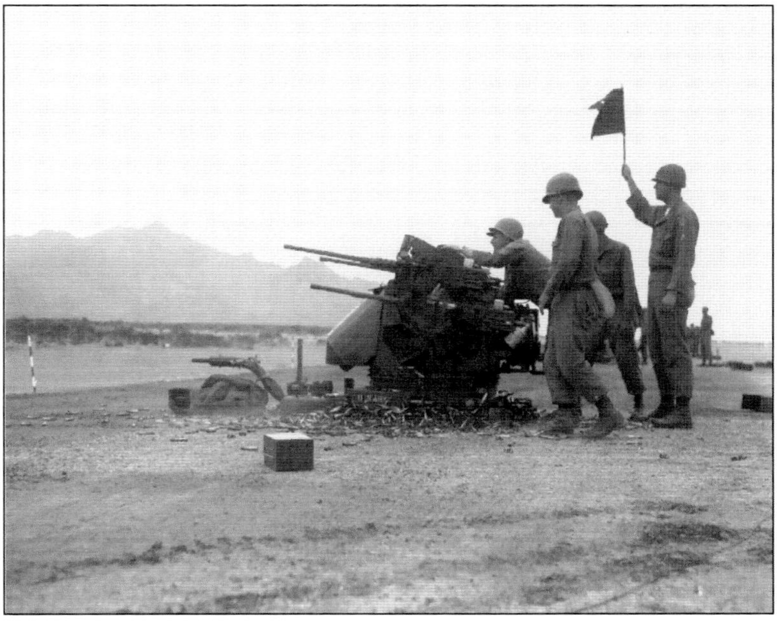

Low-flying aircraft still required guns for air defense. The venerable Quad .50 machine guns mounted four .50-caliber guns that could fire up to 2,000 rounds per minute. They were effective against propeller aircraft as well as ground targets. This crew trains on Doña Ana Range in 1954. The Army used truck-mounted Quad .50 systems through the Vietnam War.

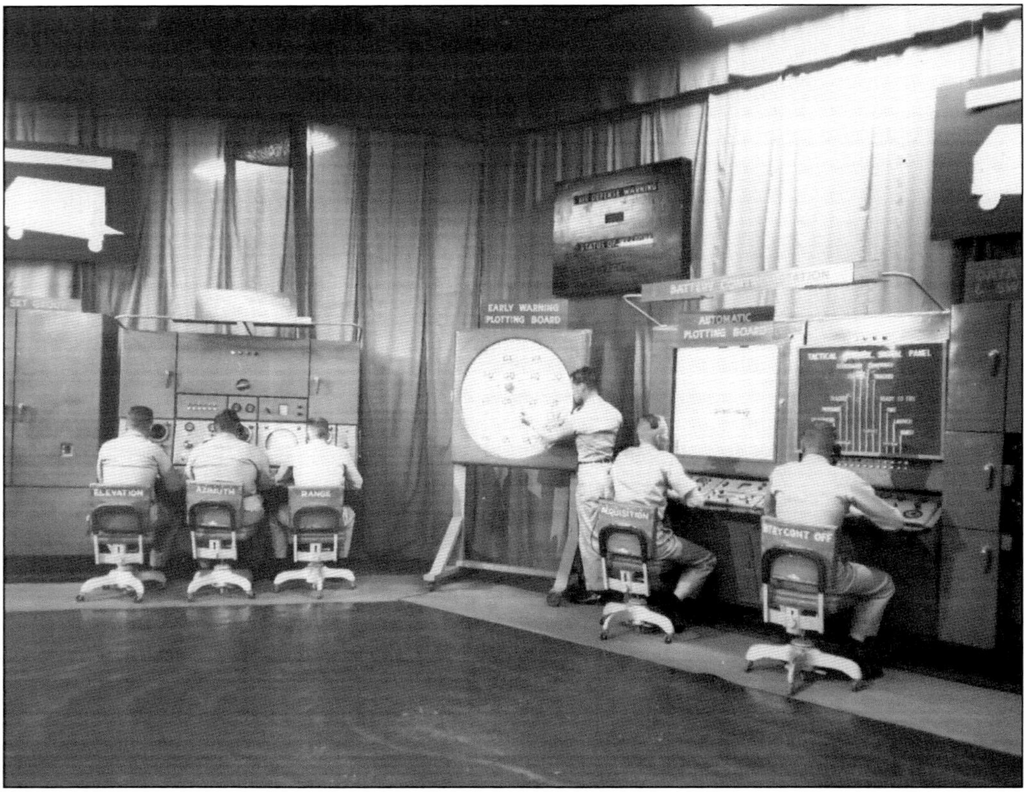

After several missile accidents near populated areas, the Army established Operation Understanding in 1957. This brought city fathers and other dignitaries to Fort Bliss to view the annual service practice for the Nike unit in their local communities. The visitors saw a portrayal of the launch procedures in Building 769. The visitors then saw an actual missile launch on the range and returned for dinner in the officers' club.

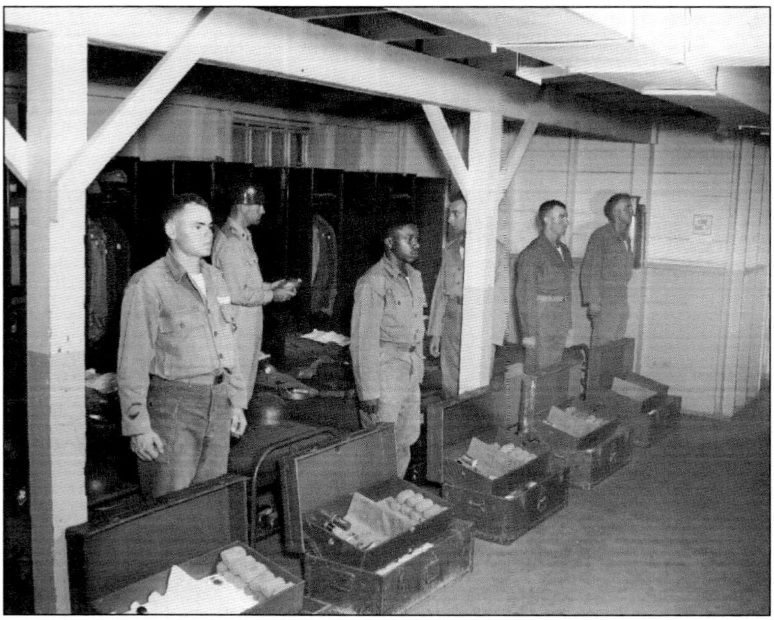

Operation of the Nike sites required trained and competent junior leaders. In 1952, the Army Air Defense Center organized a junior leader training course at Fort Bliss. This filled a dire need for leaders during and after the Korean War. Students were given actual practice in leading all aspects in antiaircraft firing, infantry combat, athletics, and other endeavors.

The leaders' course had much field training, including a leader reaction course. Student leaders were rotated through command positions in various small-unit combat scenarios. Here, students capture a prisoner in a training exercise on Castner Range in 1952.

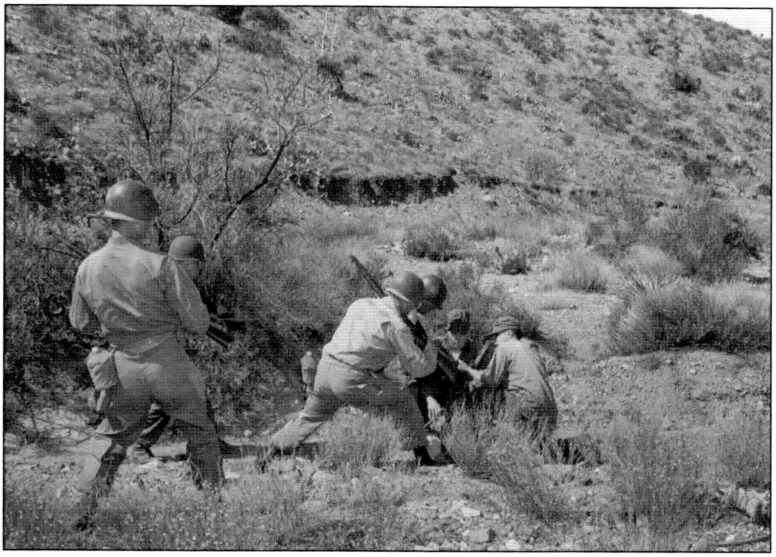

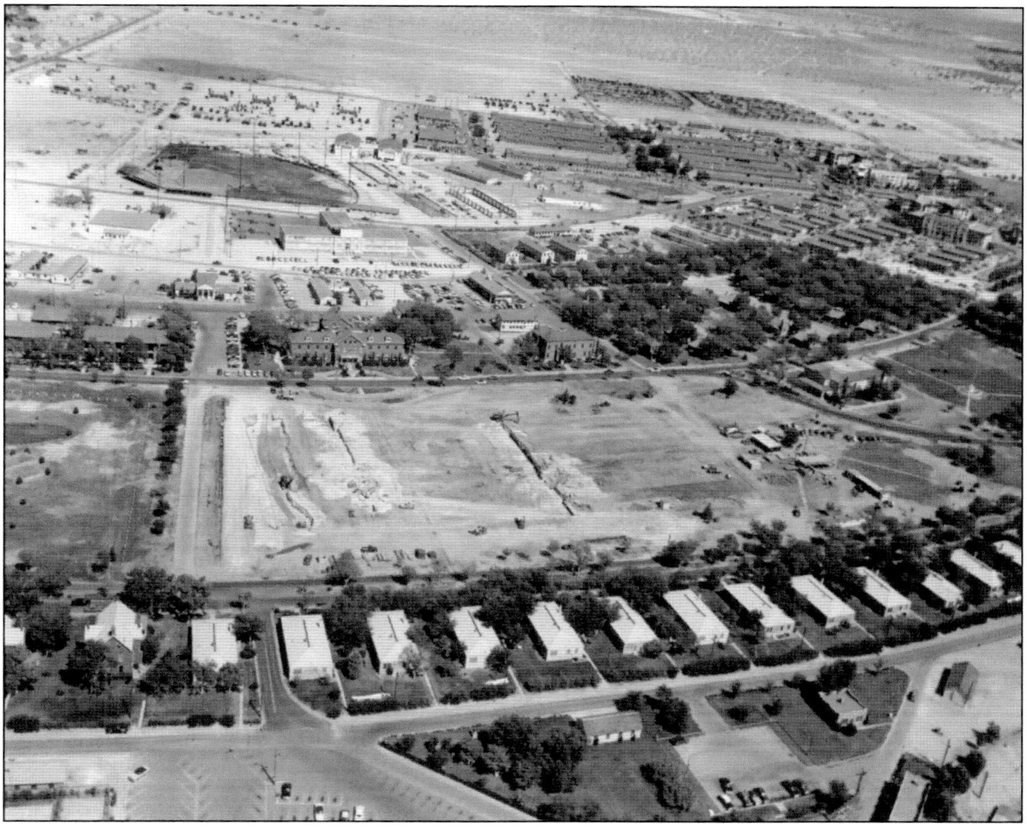

For years after World War II, the Air Defense Center struggled to train troops in the old wooden buildings hastily constructed for the war. By 1953, so many courses were taught that the school needed a new training facility. Construction on a new building started that year, between the 1893 hospital and officers' quarters. This aerial view shows that construction.

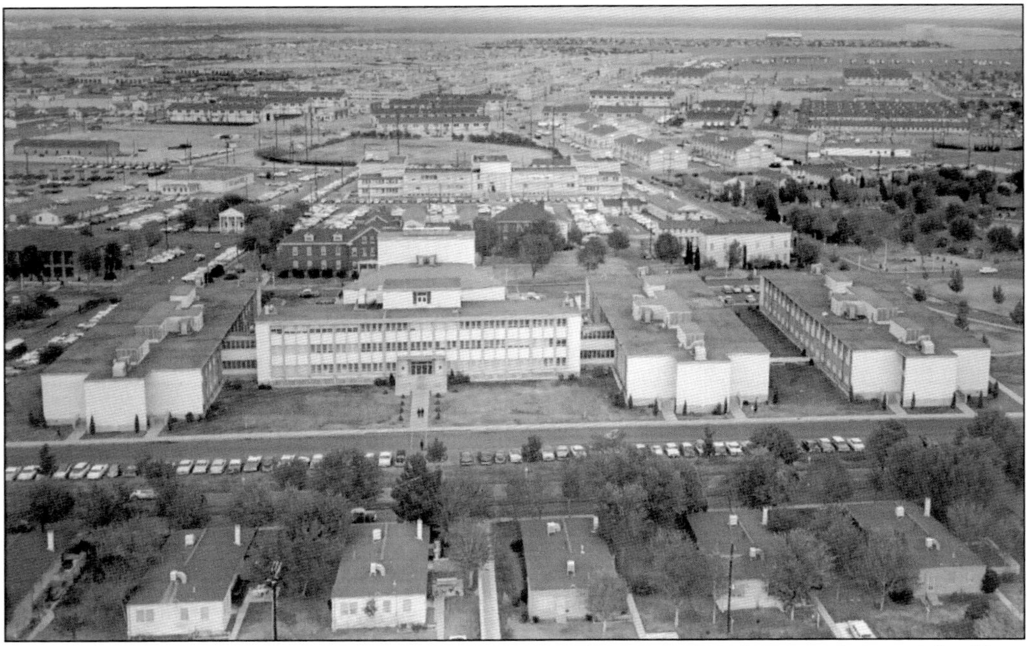

Building 2, Hinman Hall, opened on Tuesday, November 23, 1954. The sprawling new building housed the major air defense school departments, including command and staff, gunnery and materiel, electronics and engineering, and guided missiles. The building had a commander's suite on the third floor, plus a fine auditorium for seating up to 1,500 people.

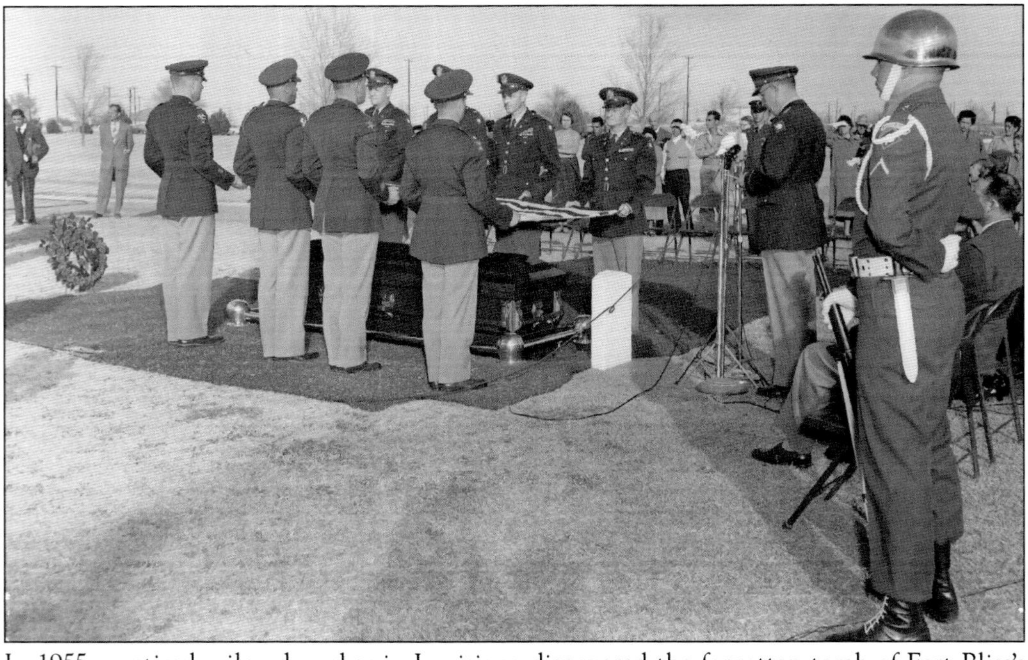

In 1955, a retired railroad worker in Louisiana discovered the forgotten tomb of Fort Bliss's namesake, Lt. Col. William W.S. Bliss. The post commander directed that units returning from Operation Sagebrush maneuvers in Louisiana recover Bliss's remains and bring them to the post. On November 22, 1955, Bliss was reinterred at the Fort Bliss National Cemetery with full military honors.

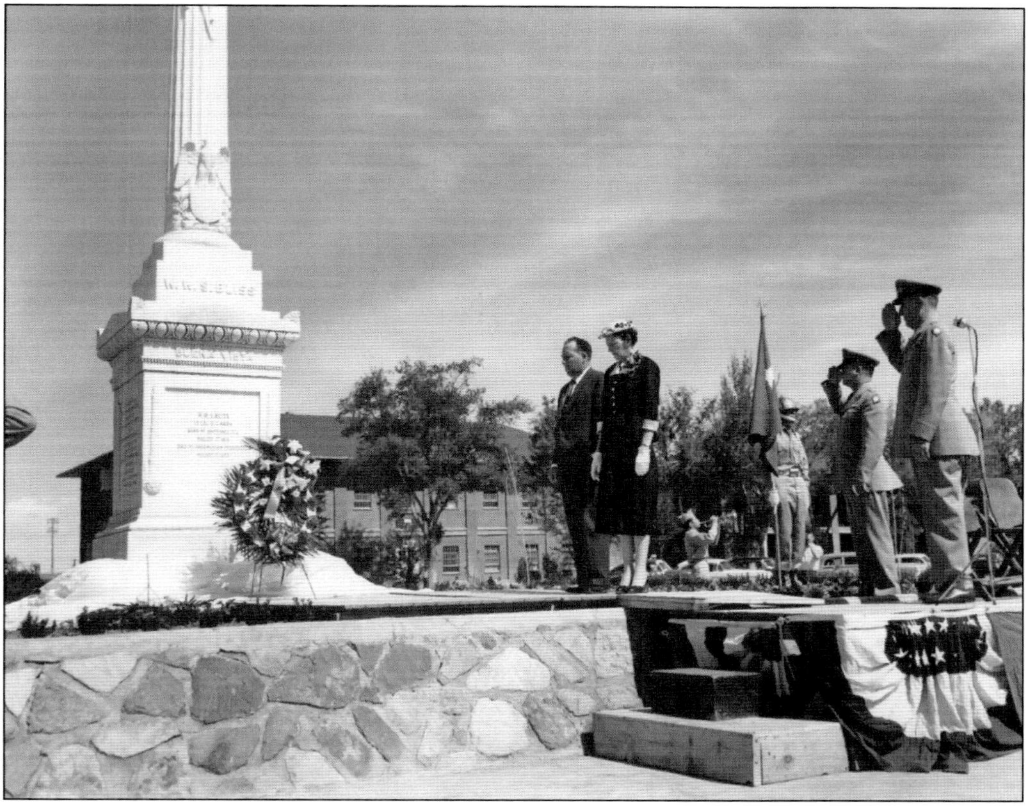

The monument over the Bliss tomb in the Girod Street Cemetery in New Orleans was also brought to Fort Bliss. It was rededicated at Smith Bliss Field on April 22, 1955, where it stands today. Mayor Pat Rodgers, of El Paso, and Margaret S. Rutledge lay a wreath. The Fort Bliss commander, Maj. Gen. Paul Rutledge, salutes at right.

With the establishment of Hinman Hall as the post headquarters in 1955, the old 1893 post headquarters received a new tenant. The Army Post Exchange (PX) moved into the building and was opened for business in 1957. This served as the Army PX until a larger facility opened on Marshall Road.

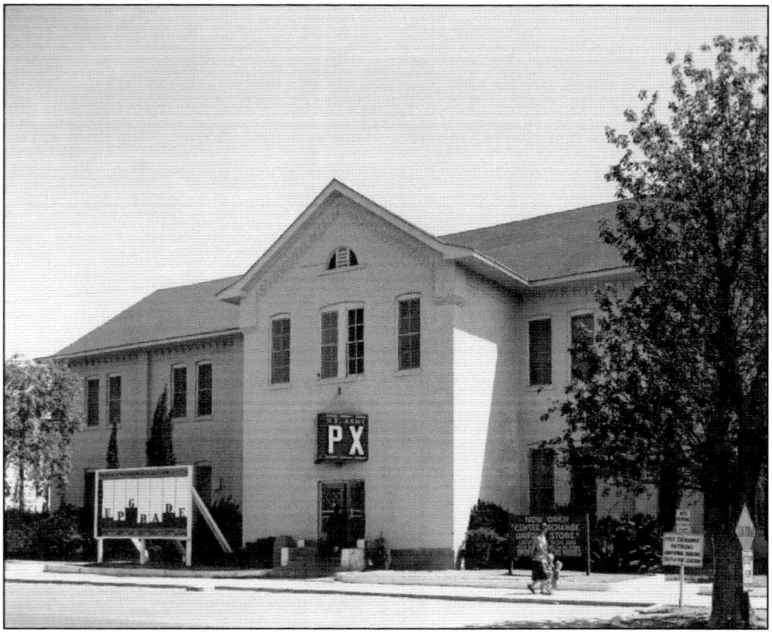

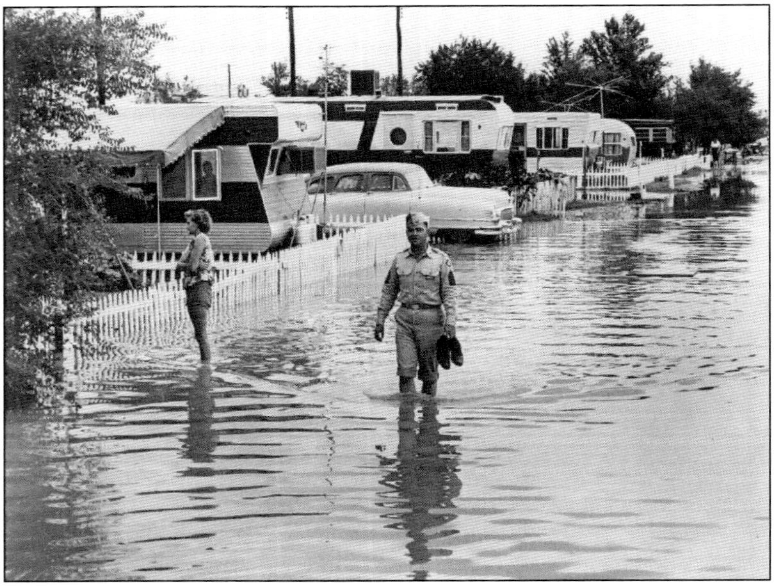

There were several family housing crises on Fort Bliss in the 1940s and 1950s. One solution was to open a trailer park, as shown here. By 1949, there were enough spaces to support 200 families. But El Paso's clear climate was occasionally marred by rainstorms and flash floods. Here, a weary sergeant wades home after a hard day.

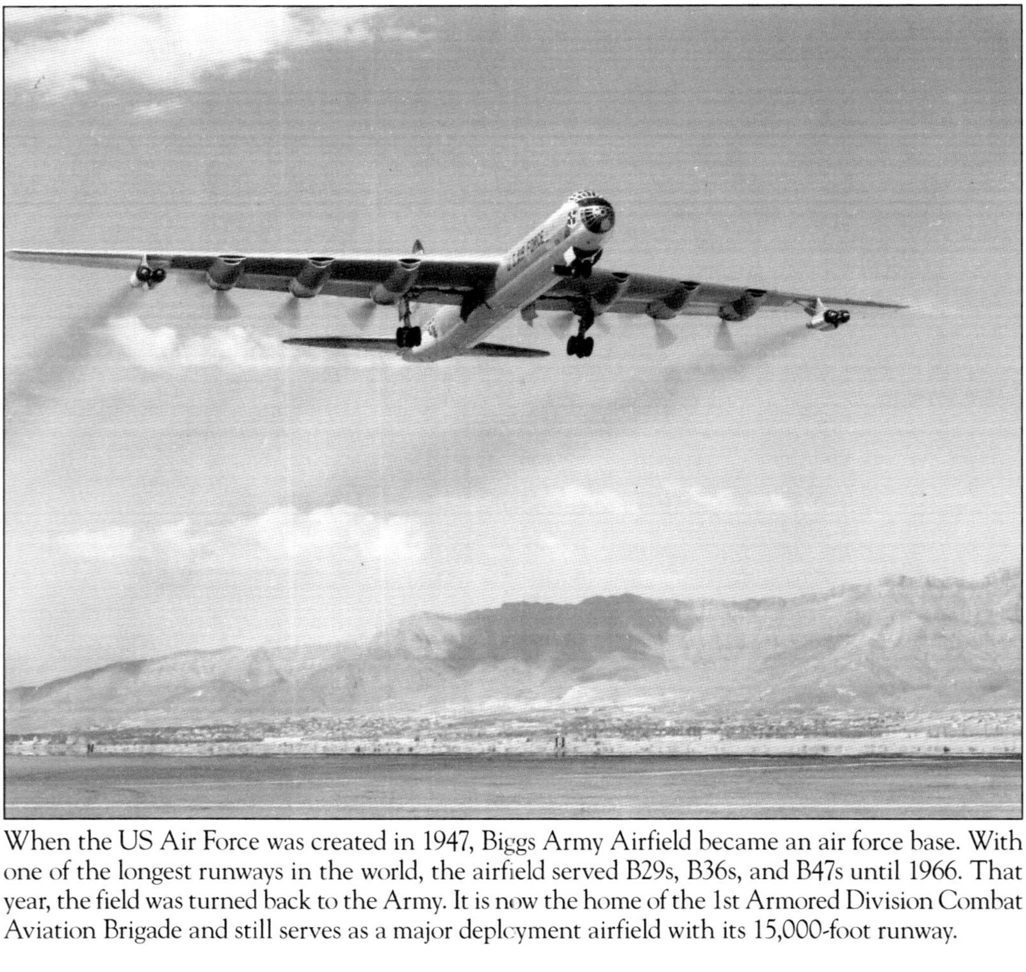

When the US Air Force was created in 1947, Biggs Army Airfield became an air force base. With one of the longest runways in the world, the airfield served B29s, B36s, and B47s until 1966. That year, the field was turned back to the Army. It is now the home of the 1st Armored Division Combat Aviation Brigade and still serves as a major deployment airfield with its 15,000-foot runway.

Army basic training ran several times on Fort Bliss: from 1950 to 1955, 1956 to 1957, 1967 to 1969, and from 1975 to 1990. Here, a sergeant instructs a recruit on the way to attain a proper seal for a chemical protective mask.

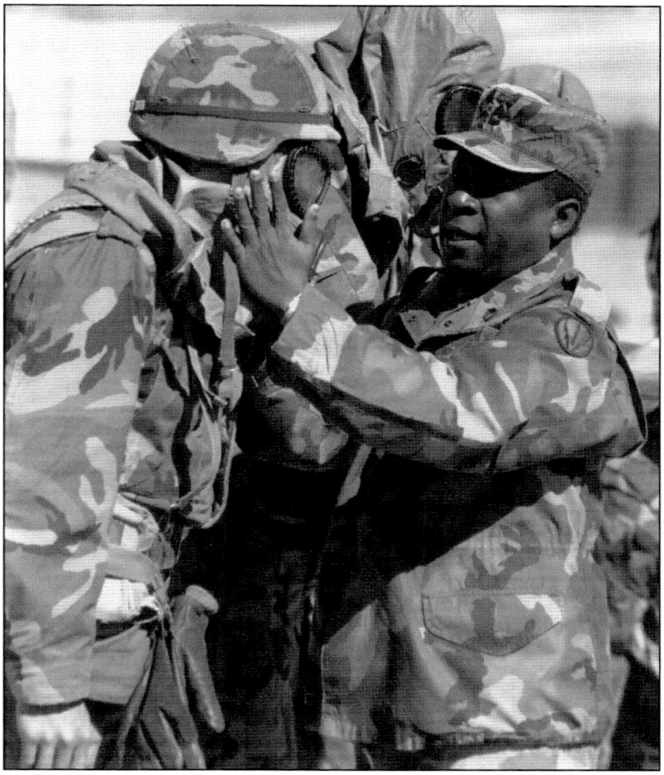

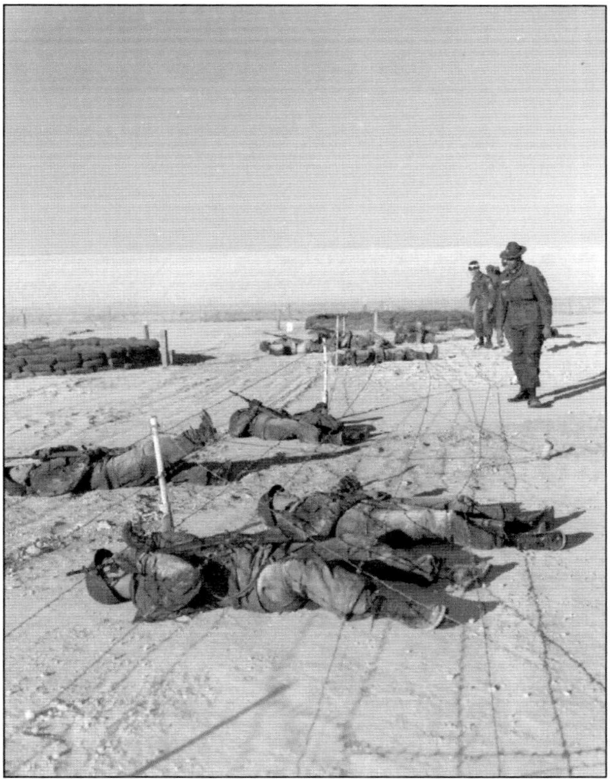

Basic trainees underwent demanding training to prepare them for Army service. Here, they negotiate the infiltration course under the watchful eyes of their drill sergeants. They are armed with M14 rifles and dressed for cold weather. The sandbags enclose explosive charges that were detonated during the exercise. After a practice run, soldiers repeated this with explosions and live machine gun fire overhead. No one stood up on the live course.

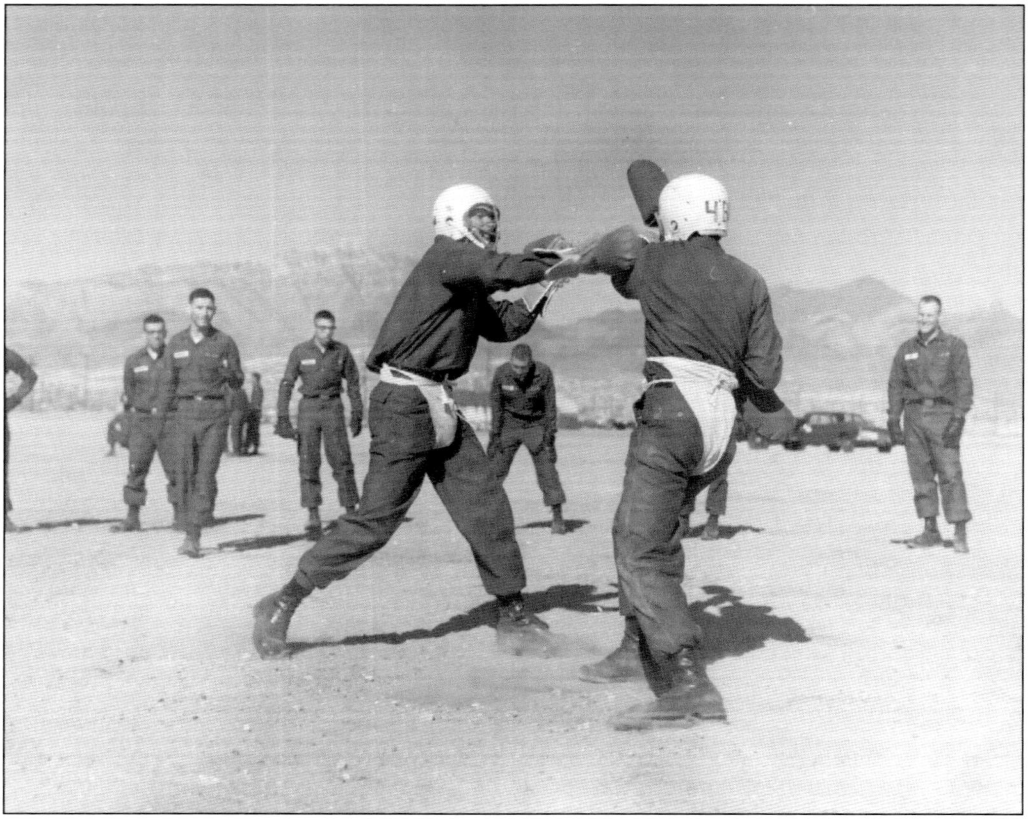

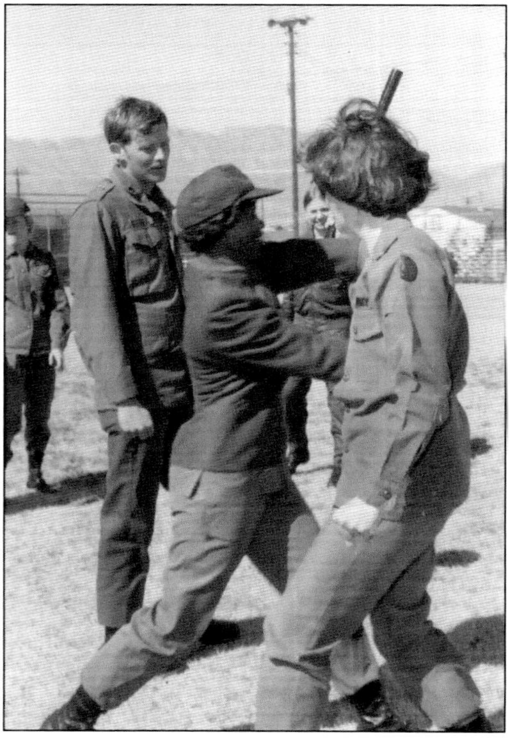

Pugil stick training served as a safe way to teach aggressive bayonet training. The trainees wore protective equipment and sparred with staffs padded at both ends. Even so, troops could be injured during this training. Both the US Army and the Marine Corps still do pugil stick training.

Female soldiers undergo hand-to-hand combat training at Fort Bliss. The Women's Army Corps was eliminated in 1978 when female soldiers were integrated into the existing Army branches. Although they were prohibited from direct combat, they still trained on weapons and in combatives. Today, all Army job series are open to females.

Armed Forces Day was a day to display the Army's weapons systems to the public. The post held exhibitions on Smith Bliss Field, Noel Parade Field, and Biggs Army Airfield. Here, a soldier explains a Hawk Missile System to a visitor.

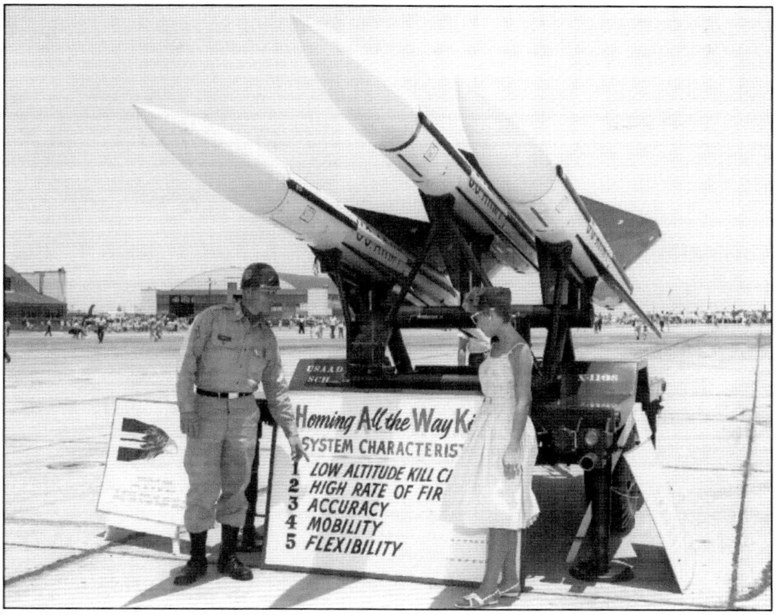

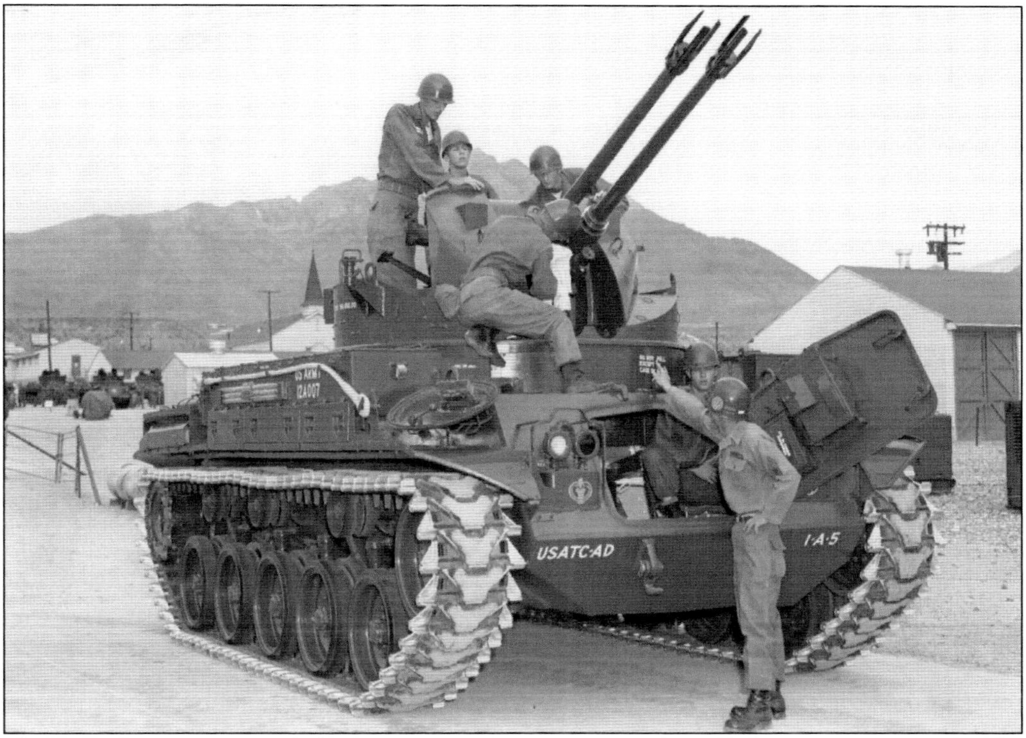

The M42 twin 40-mm Duster system was used in Vietnam and was a critical element in the Battle of Hue during 1968. These troops trained on the system before deploying. Lacking these systems in the active Army inventory, the Army drew them from the Army National Guard for use in Vietnam.

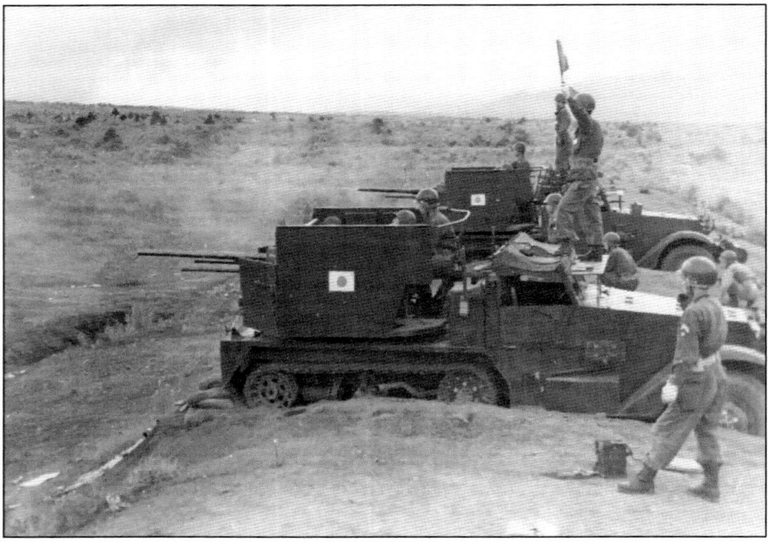

In the 1950s, Fort Bliss began to train allied soldiers on US weapons systems. Here, Japanese troops train on the M15 half-track. They are on a range at Doña Ana, controlled by their own troops. Japanese troops still conduct live fire training on McGregor Range every year.

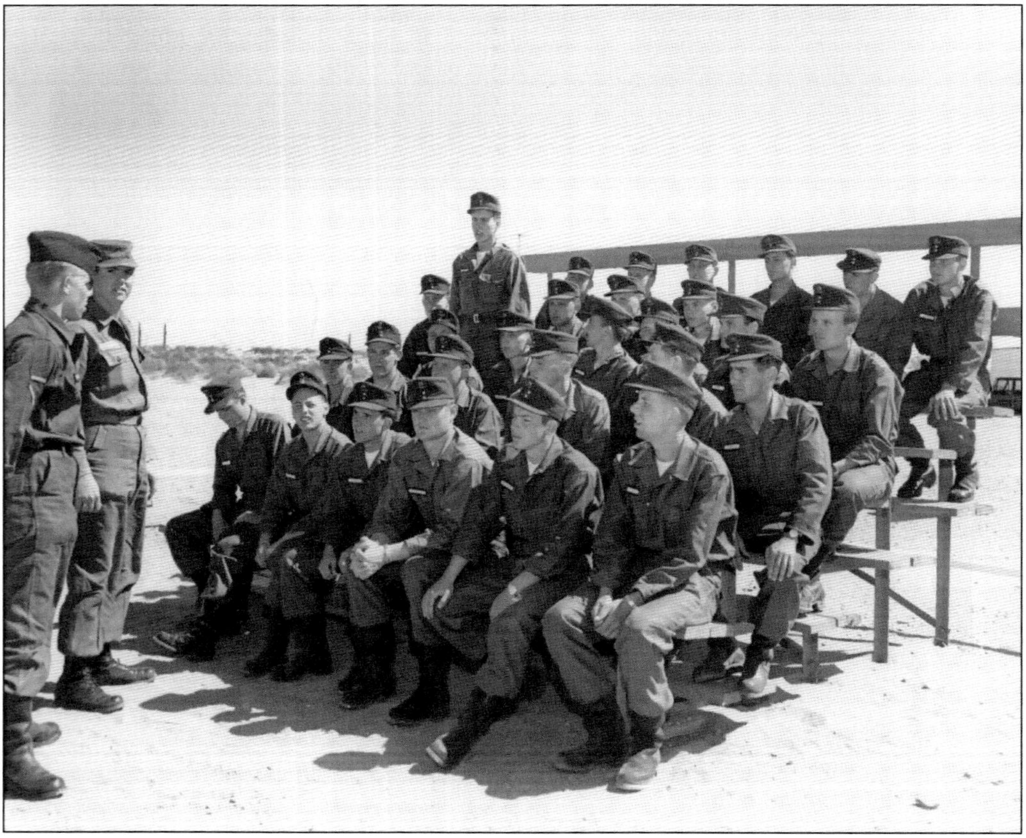

By 1956, allied soldiers were undergoing air defense training at Fort Bliss. In 1957, the air defense center established a liaison office to oversee German air defense training. Here, Sgt. Earl Phillips, of the 1st Guided Missile Battalion, introduces the Nike missile to German Luftwaffe soldiers. In 1966, the German Air Force Air Defense School opened. Since then, more than 45,000 German soldiers have trained here.

Lacking a small manpacked air defense system, the Army fielded the Redeye in 1968. Training for Redeye crewmen began at Fort Bliss in 1966. This was revolutionary; it was the first time a soldier could carry an air defense weapon on his or her person.

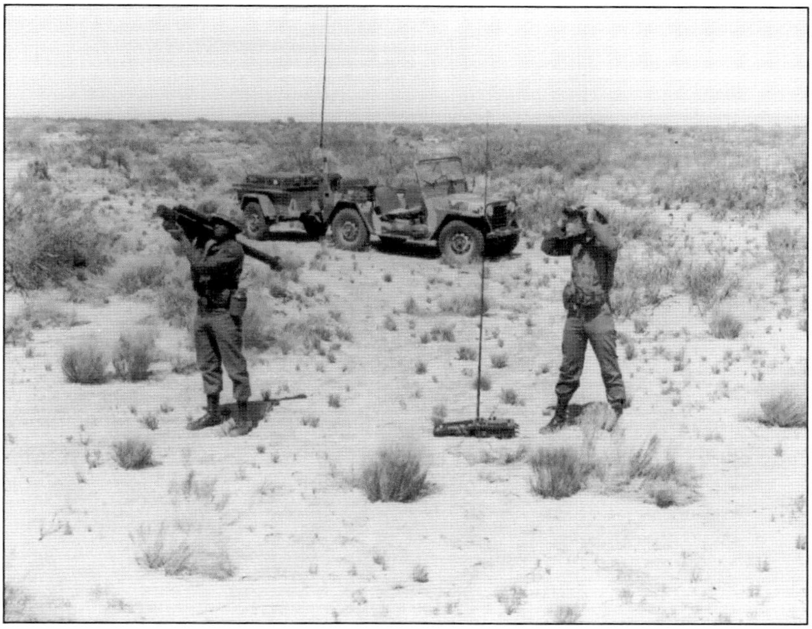

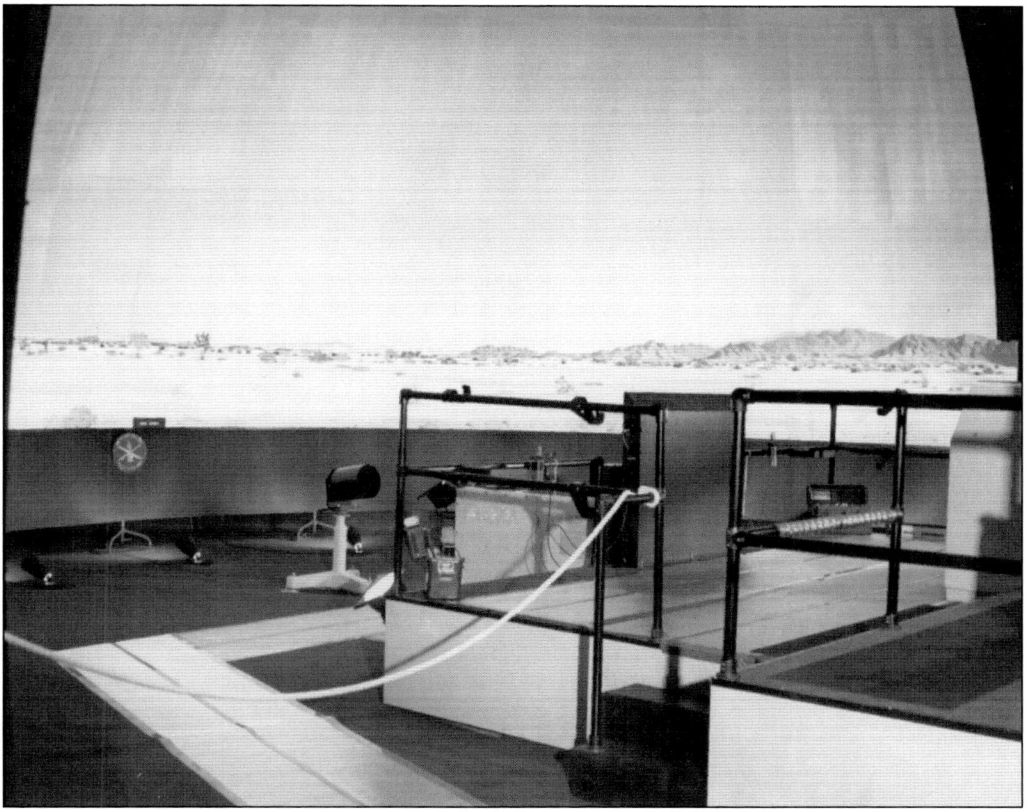

The interior of the Redeye missile moving target simulator was a one-quarter hemisphere. The horizon appeared on the projection, with air space above it. Projections of aircraft could be shown approaching from any direction at any altitude. A Redeye missile simulator could acquire a target and simulate a launch. The Army built three of these on the post.

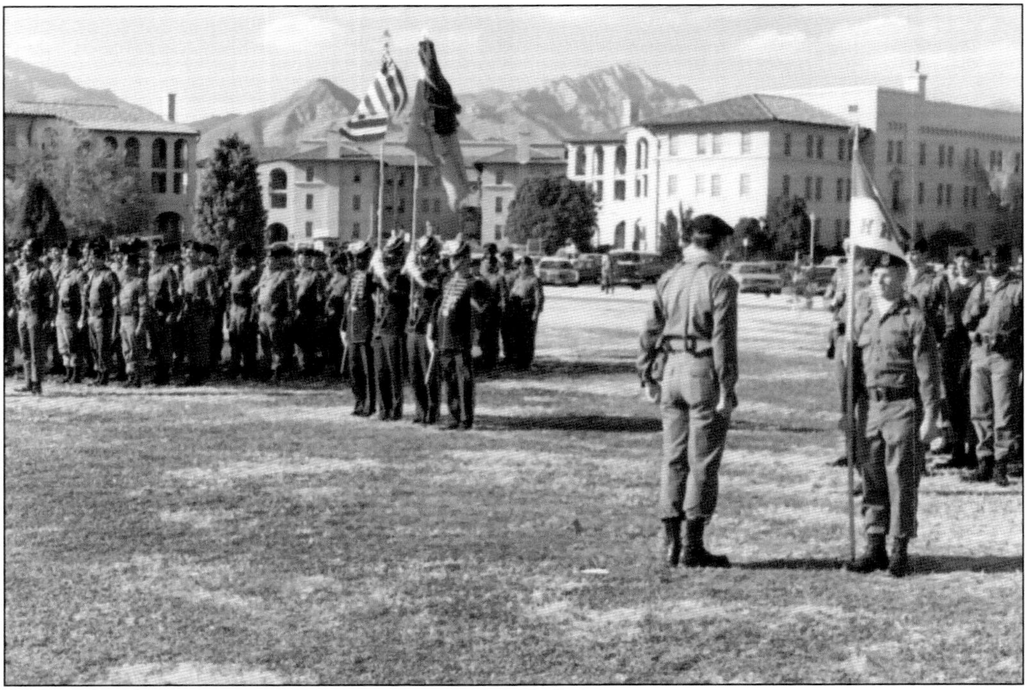

After a 112-year absence, the 3rd Armored Cavalry Regiment returned to Fort Bliss in 1972. This was the first time in 29 years that Fort Bliss hosted cavalry troops. This time, they were fully mechanized, with not a horse to be seen. (Courtesy 3rd Armored Cavalry Regiment Museum.)

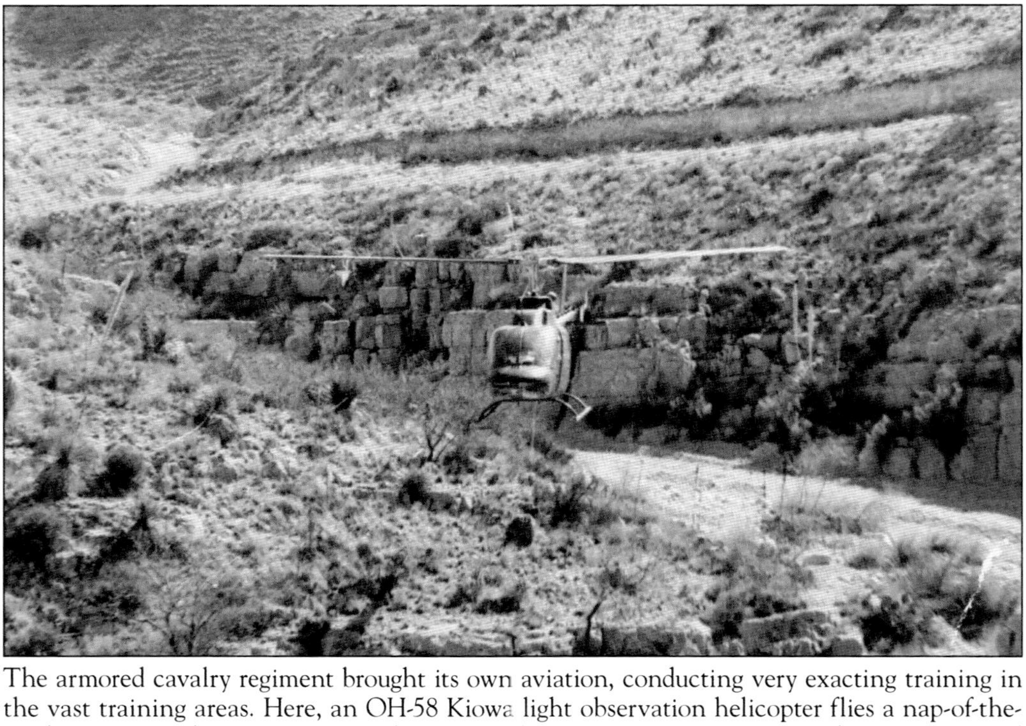

The armored cavalry regiment brought its own aviation, conducting very exacting training in the vast training areas. Here, an OH-58 Kiowa light observation helicopter flies a nap-of-the-earth mission in the training area. This was a risky maneuver, requiring many hours to gain and maintain proficiency. (Courtesy 3rd Armored Cavalry Regiment Museum.)

By 1969, the post population had again outgrown its hospital. Accordingly, the Army authorized construction of a new facility. Building began in earnest in 1970, with a new multistory hospital going up at the west end of the hospital complex.

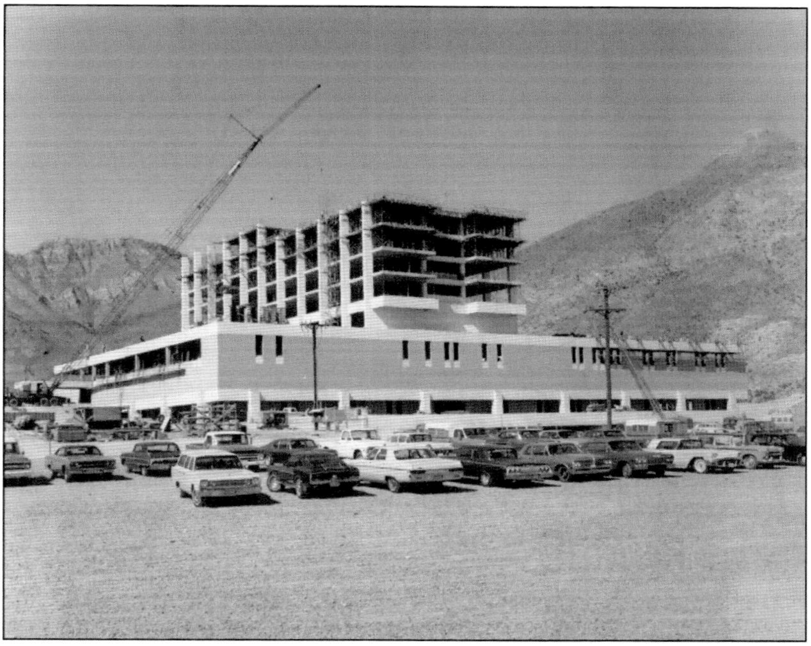

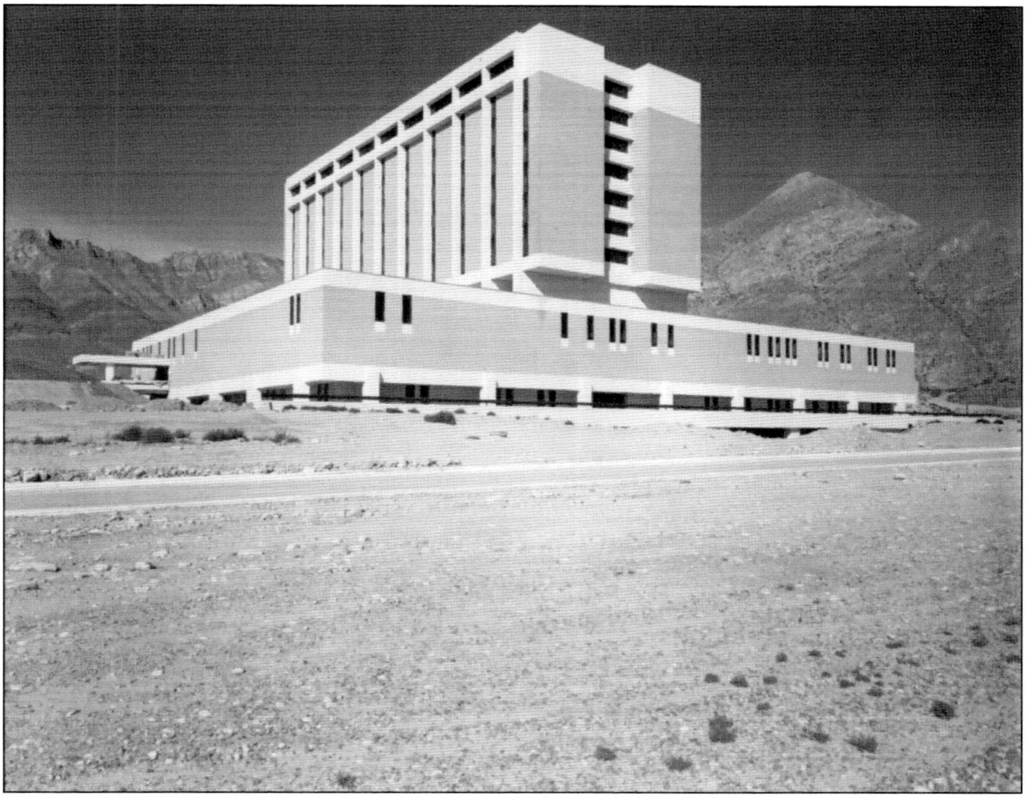

By 1972, the new William Beaumont General Hospital opened for patients. The new facility provided every possible modern medical and surgical service to active and retired service members. Many of the old buildings in the hospital area have been vacated and demolished, but some of them remain and are in use. This facility will soon be replaced by a new hospital to the east.

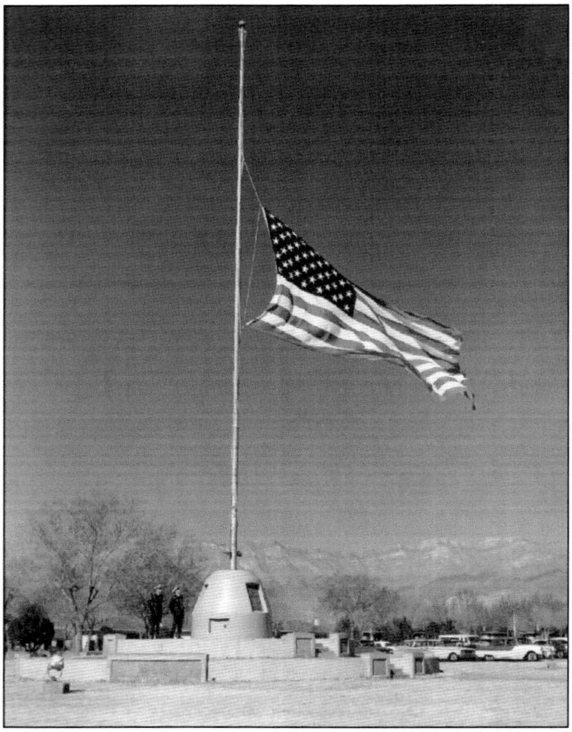

The post flag flies at half-mast on Memorial Circle at Fort Bliss. Since the arrival of the 1st Armored Division in 2011, the historical west side remains and is still used. The 1893 quarters are still occupied by senior officers. The barracks built in 1893 and after are still used for administrative purposes.

A final retreat ceremony ends the duty day in 1959. The kids in the picture are now grown, and the military police soldiers are old men if they are still alive. The post will celebrate its 200 year anniversary in 2048. One wonders what the post will look like then.

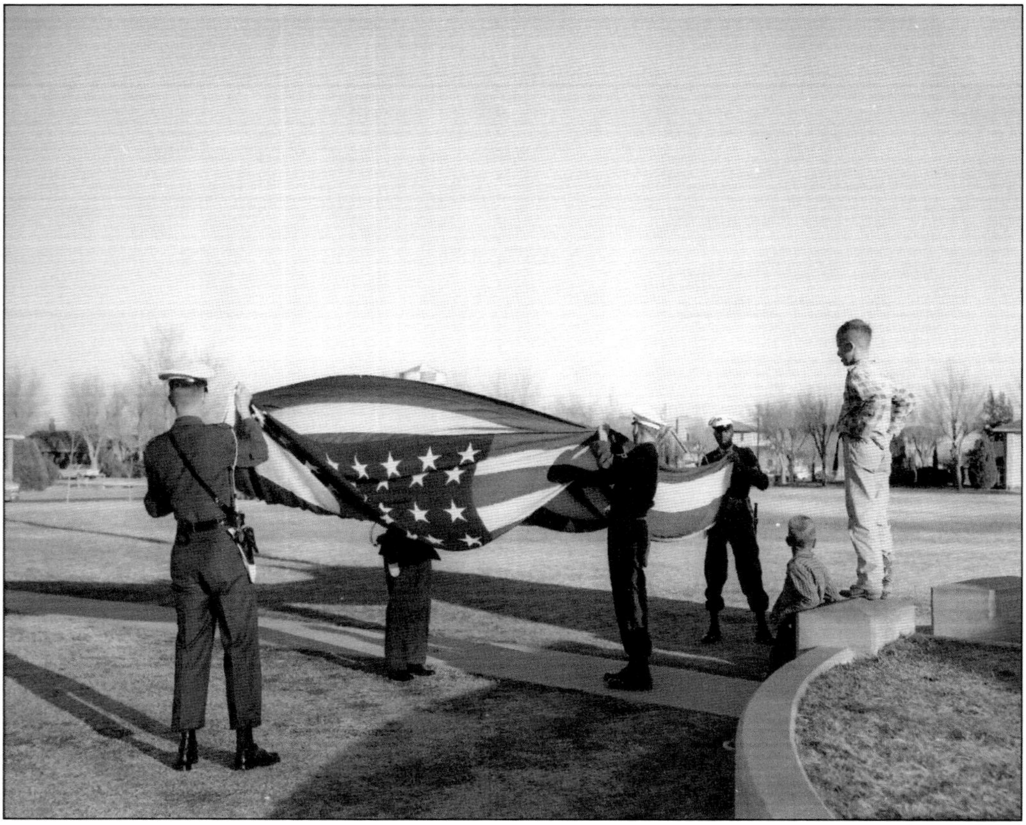

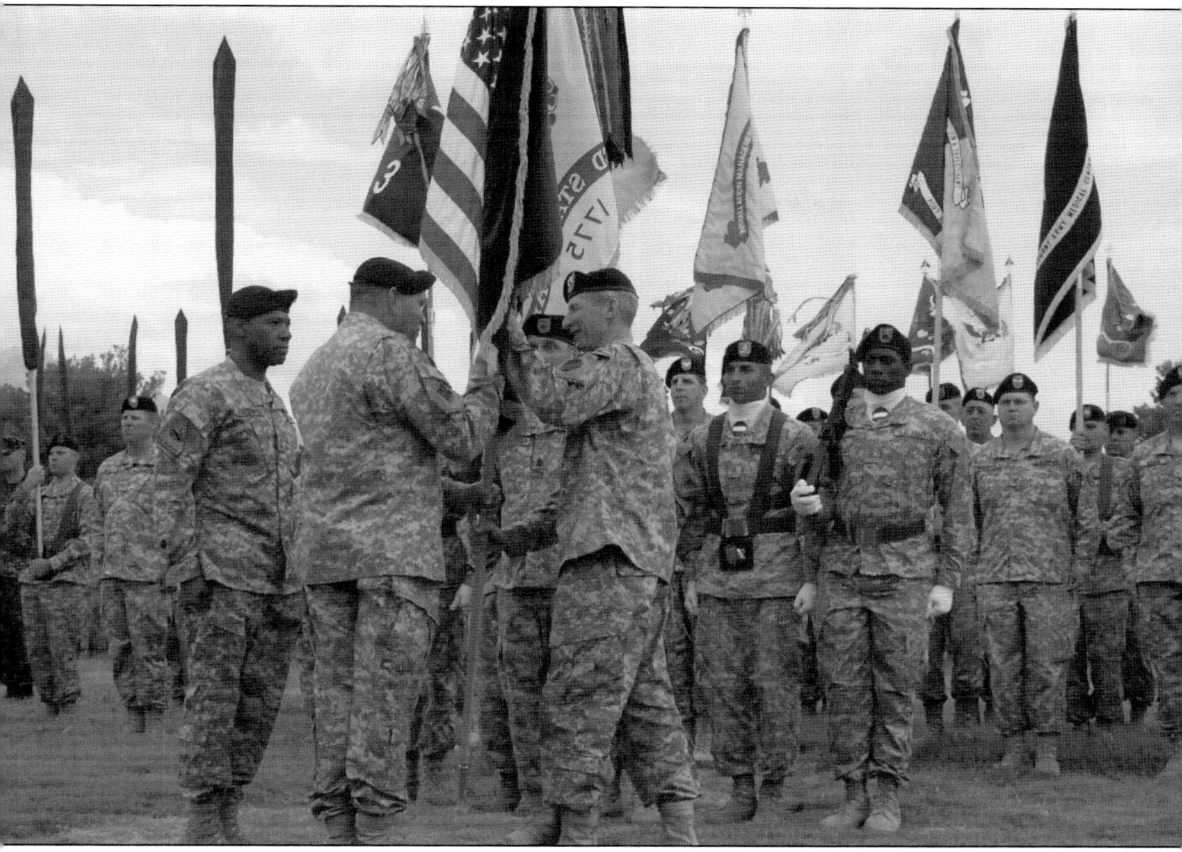

The command of the post changed from an air defense post to a major maneuver center in 2011. Here, Maj. Gen. Howard Bromberg presents the post colors to Maj. Gen. Dana J.H. Pittard. The post now hosts combined arms combat power and serves as a major Army training and deployment center. The long history of Fort Bliss continues.

Bibliography

Carson, James. *Against the Grain, Colonel Henry M. Lazelle and the U.S. Army*. Denton, TX: UNT Press, 2015.

Crego, Arthur Van Voorhis. *City on The Mesa, The New Fort Bliss 1890 to 1895*. Fort Bliss, TX: Fort Bliss Museums Division, DPTMS, 1993 (Reprint).

Hamilton, John A. *Blazing Skies, Air Defense Artillery on Fort Bliss, Texas, 1940–2009*. Washington, DC: US Government Printing Office, 2009.

Harris III, Charles H., and Louis R. Sadler. *The Secret War in El Paso, Mexican Revolutionary Intrigue, 1906–1920*. Albuquerque, NM: University of New Mexico Press, 2009.

Jamieson, Perry. *A Survey History of Fort Bliss, 1890–1940*. Fort Bliss, TX: Cultural Resources Management Division, Directorate of the Environment, US Army Air Defense Artillery Center, 1993.

Lane, Lydia Spencer. *I Married a Soldier, or Old Days in The Old Army*. Albuquerque, NM: Horn & Wallace, Publishers, Inc., 1964.

McMaster, Richard K. *Musket, Saber, and Missile, A History of Fort Bliss*. El Paso, TX: Complete Printing & Letter Service, 1962, 1963.

Metz, Leon C. *Fort Bliss, An Illustrated History*. El Paso, TX: Mangan Books, 1981.

———. *Desert Army, Fort Bliss on The Texas Border*. El Paso, TX: Mangan Books, 1988.

Sonnichsen, C.L. *Pass of The North, Four Centuries on the Rio Grande, Vols. I and II*. El Paso, TX: Texas Western Press, 1968 (Vol. I) and 1980 (Vol. II).

The Story of Fort Bliss. Fort Bliss, TX: US Army Air Defense Artillery Center, 1966.

Thomlinson, M.H. *The Garrison of Fort Bliss, 1849–1916*. El Paso, TX: Hertzog & Resler, Printers, 1945.

Timmons, W.H. *James Wiley Magoffin, Don Santiago—Texas Pioneer*. El Paso, TX: Texas Western Press, 1999.

Wilson, John B. *Maneuver and Firepower, The Evolution of Divisions and Separate Brigades*. Washington, DC: US Army Center of Military History, 1997.

The 1st Armored Division and Fort Bliss Museum

The 1st Armored Division and Fort Bliss Museum is located on Fort Bliss, Texas, in Building 1735, Marshall Road, just north of the Freedom Crossing shopping complex. The mission of the museum is to serve the soldiers their family members, Army civilians and contractors, and the local El Paso community by presenting the history and heritage of Fort Bliss and its major tenant units. The museum features the history of the post dating from its establishment in 1849; its six locations in and around El Paso, Texas; and its transformation from one of many frontier garrisons to the major maneuver center it is today. It houses many artifacts from its history, including frontier wagons and equipment, the many different uniforms and equipment carried by soldiers over the years, and large vehicles such as tanks and aircraft. The museum is open from Monday through Friday from 9:00 a.m. to 4:00 p.m. and on Saturday from 10:00 a.m. to 3:00 p.m. It is closed on federal holidays. Admission is free.